D0571487

A BOX OF GANDYS

(WITH HARRIET'S COMPLIMENTS)

THE
BIGGLE BERRY BOOK

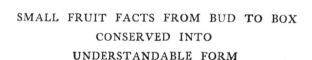

SMALL FRUIT FACTS FROM BUD TO BOX
CONSERVED INTO
UNDERSTANDABLE FORM

BY

JACOB BIGGLE

ILLUSTRATED

*Fill your lap and fill your bosom; only spare the berry
blossom.* —WORDSWORTH.

Skyhorse Publishing

Skyhorse Publishing books may be purchased in bulk at special discounts for sales promotion, corporate gifts, fund-raising, or educational purposes. Special editions can also be created to specifications. For details, contact the Special Sales Department, Skyhorse Publishing, 307 West 36th Street, 11th Floor, New York, NY 10018 or info@ skyhorsepublishing.com.

Skyhorse® and Skyhorse Publishing® are registered trademarks of Skyhorse Publishing, Inc.®, a Delaware corporation.

Visit our website at www.skyhorsepublishing.com.

10 9 8 7 6 5 4 3 2 1

Library of Congress Cataloging-in-Publication Data is available on file.
ISBN: 978-1-62636-143-0

Printed in China

CONTENTS

PREFACE

I HOLD that it is right to tell what we know in any line of farming, if our knowledge be of value to others and will help them to success. Now, I have been engaged more or less in strawberry culture for many years, and have in that time learned a little, and this little I am ready to communicate to my neighbors and even to impart to a wider circle, wide enough to take in the whole Farm Journal family and the entire remnant of the population of the country.

The only trouble is, I do not know it all ; and yet it may be best that I do not, since I have discovered that those folks who know it all are apt to get behind the lighthouse and are left in the dark themselves.

Confessed, I do not know it all ; yet Harriet knows some and Tim knows a heap ; together, though, we are so far from a universal knowing that I have not hesitated, in preparing this book for publication, to get the opinions and experiences of a number of bright, practical men.

It will be seen, therefore, that some pages of my book will contain explicit information from other berry growers,—living in all parts of the country, in all latitudes and longitudes. In brief, I have tried to make the book national in its scope, rather than local. And I herewith extend my cordial thanks for the outside information which has enabled me to do so.

A feature is the showing of the berries in natural colors, which, to my knowledge, has never before been successfully accomplished in a book. It cost time, money and infinite pains to procure accurate paintings of the fruits and to transfer them to the pages of this book, many specimens being printed in eight separate colors in order to produce the required truthfulness of shading. Of course most of the credit of success in this line must accrue to the publishers, and to them I freely give it.

Another feature is the many excellent half-tone engravings which were made, from photographs, expressly for this book. These photographs came from all parts of this great country of ours, and show actual scenes, appliances, methods, etc.

Grapes, although perhaps not strictly in the classification of small fruits, are given a chapter (or may I call it an arbor?) all by themselves; for, surely, a fruit garden without grape-vines would be like a pudding minus sauce.

My earnest wish is that this little book may lead its readers far into that place of delight—the finished fruit-garden.

JACOB BIGGLE.

Elmwood,
1911.

ILLUSTRATIONS IN COLOR

CHAPTER I

MAKING A BEGINNING

The way to begin is to begin

Among the enthusiastic growers, whose opinions about berry culture I have asked, is J. H. Hale, of the state of Connecticut, and the United States of America, for he belongs to the latter; and here is one of the things he wrote : "No man should fool himself into telling his wife that he hasn't time to bother with such small trash as berries, but will buy all the family wants ; he may not be much of a liar, but those of us who have so often heard that old chestnut about 'buying all the berries the family wants,' know that man is 'way off.' He never did and never will buy one-tenth part as many berries as the family will consume, if he will give them all they can wallow in right fresh from the home garden." Mr. Hale is right ; few in the country will buy enough berries.

The only just and true way for an honorable and manly man is to grow them, and let everybody about the place have all he can eat. Then there'll be less lard, tough beef, or dried-apple pies to be manipulated and cooked in midsummer over red-hot ranges. For the berry comes from the garden to the table in tempting and presentable shape, fit to grace the table of a king.

A friend asks : "How many berries will the average farmer buy ? Will it be one quart a week ?" A housewife was promised by her well-to-do husband, that instead of growing berries he would purchase all she wanted. At the end of the season she said : "How many berries do you suppose we bought ? Not a single quart !"

That forcible question and answer are altogether too common. Farmers who with very little expense can grow these most healthful and delicious fruits, deny to themselves and their families the greatest table luxury which Providence has bestowed upon people of temperate climates, when a single square rod of ground might yield them more intrinsic value than an acre in many other products.

Berry growing is to many people a great mystery, as the writer has had impressed upon him by numberless inquiries, both verbal and written. There is no fruit crop so immediately productive,

ROLLING CRUSHES LUMPS AND PACKS DOWN THE SOIL

none which attaches to itself so much enthusiasm and quick reward for labor expended. Berries flourish in nearly all soils and in all temperate climates. The number of varieties is now unlimited, and suited to all tastes.

One large farmer in the country consigns to his own table a peck a day ; others provide a quart for each person, and dispense almost wholly with meat so long as berries can be had in good condition. A very intelligent young lady living opposite my farm, who has traveled the world over, enjoys life just as long as the supply of berries continues; but at other seasons she is more or less of an invalid. And yet there are too many who regard berries as mere luxuries, and refer you to pork and potatoes for nourishment and substantial sustenance for body and mind.

A. I. Root says : "Everybody ought to have all the berries he wants. If he does not care to grow them, he ought to be in some business so that he can afford to buy them, quart after quart, morning, noon and night. Not only because they give enjoyment, but because they are the cheapest, best and most natural medicine to tone up the system that has ever been invented. They are both victuals and drink. The man who can not afford to give up his beer, tea and coffee, yes, and tobacco, too, when berries are plentiful and cheap, is a man to be pitied."

Then—outside of the farming class—there are thousands of town dwellers and suburbanites who might grow berries, if they would, on their little plots of ground in the back yard. Why not ? This book is for them as well as for the farmer and the professional grower ; and any one can learn who will study its teachings.

It is difficult to give accurate directions as to the selection of a location for fruits or to describe a soil that will bring the best results. There are a few general principles, however, says Gabriel Hiester, of

Pennsylvania, that have become firmly established by the experience of the most careful horticulturists extending back through the past century. There are several points to be considered in selecting a location, which apply to all fruits, and they may be briefly stated, as follows :

SOIL.—All fruits do best on a *deep* soil on an open subsoil that will allow perfect drainage. Let me then impress upon the mind of the reader that the first requisite for the profitable production of fruit of any kind is an *open subsoil* that will allow perfect

FOLLOW THE DISC HARROW WITH A SMOOTHING HARROW

drainage; second a *deep top-soil* of a character suited to the kind of fruit grown. (The character of soil best suited to each kind of fruit will be treated later on in this book.)

EXPOSURE. — Opinions differ somewhat on this point, but a majority seem to favor a northern exposure, as the idea prevails that the buds are there retarded somewhat, and are less liable to be injured by late frosts. Also, that a southern exposure is least desirable, except when extra-early fruit is wanted.

ALTITUDE.—This is more important than exposure. Fruit should be planted above the level of the "lake" of cold air that settles in the valleys at night ; the warmer and more sheltered the valley the more important is this point, as these places are most subject to late frosts. No fixed height can be given at which it will be safe to plant ; it will depend upon the width of the valley and the abruptness of the slope at either side. Each planter must decide for himself what will be a safe altitude for protection against untimely spring frosts.

Conditions differ, however, along the shores of lakes and broad rivers ; here the *water* tempers the air and prevents injury by late frosts. For instance, the influence of Lake Erie extends two or three miles inland; and along the Susquehanna, fruit which grows close to the bank frequently escapes injury, while a mile back in the country the entire crop is destroyed by frost. Good fruit can be grown on these bottom lands which border on rivers and lakes, provided they are well drained either naturally or artificially.

RAINFALL.—The distribution of rainfall throughout the entire season is important. Fruit requires a large amount of moisture during the growing season, and unless this is supplied by rains, you will have to practise irrigation. (See Chapter II.)

ARTIFICIAL DRAINAGE.—If the ground is not naturally well drained, it will pay to install a system of underground tile drains. This, however, is too intricate a subject to be thoroughly treated in the space at my disposal ; therefore, I would suggest that you write to the Secretary, U. S. Department of Agriculture, Washington, D. C., and ask for a copy of Farmers' Bulletin No. 187, entitled "Drainage of Land." And remember this : "Tile drains help

THESE NICE BERRIES CAME FROM A GOOD BEGINNING

to make wet soils drier, and dry soils more moist.'' Harriet says that this sounds like a paradox, but I can earnestly assert that experience has proved it to be a fact.

SHIPPING OR MARKETING FACILITIES.—After fruit is grown, it must, in many instances, be marketed. Therefore, the prospective grower needs to take into consideration, when selecting land, the distance from the railway station or steamer wharf, or the hauling distance to stores or customers. Good roads, or competing lines of railways (which usually mean cheaper freights), are both important factors to consider. (For additional marketing suggestions, consult Chapter XIV.)

PRELIMINARY PREPARATION.—Nearly all grow-ers wisely recommend preparing the ground a year or two *before* the berries are to be planted, by planting other crops which must be cultivated, and thus getting the soil mellow and the weed seeds sprouted and out of the way. A one-year *clover* sod, well manured and planted to potatoes and well tilled one year, makes one of the best preparations for berries ; but any other plan that will make the soil reasonably rich and in good tilth and free from weed seeds, will answer.

Never plant small fruits on land which has been in old *grass* sod, until it has been cultivated two or three seasons. Why ? Because such land is apt to be full of destructive ''white grubs,'' which are mostly eradicated by cultivation.

FERTILITY AND MANURING.—This is an intricate subject, and I lack space to treat it thoroughly here.

For full details on this topic I suggest that you read Chapter IV of the Biggle Garden Book (published uniform with this volume), and write to the Secretary, U. S. Department of Agriculture, Washington, D. C., and ask for Farmers' Bulletins Nos. 44, 192, 245, 257 and 278. In this way you will learn all about nitrogen, phosphoric acid, potash, nitrate of soda,

IF MANURE IS APPLIED IN FURROWS, BE SURE IT IS WELL
MIXED WITH SOIL BEFORE PLANTING

lime, humus, cover crops, nitrogenous crops, liquid manure, etc.

Generally speaking, I will say that there is no better all-purpose fertilizer than stable manure. Haul it on the ground in winter and early spring, and spread it as it is hauled ; plow it under, and then broadcast (to each acre) about 400 pounds of kainit (a commercial form of potash), and about 600 pounds of finely-ground bone meal ; harrow this in, and you have a very good, complete mixture which contains all essential elements of plant food. Or, if you desire, you can substitute muriate or sulphate of potash for the kainit, or twenty-five bushels of unleached hardwood ashes ; or phosphates or superphosphates may be substituted for the bone. If stable manure can not be obtained, and if there is sufficient humus (decayed vegetable matter) in the soil, buy a high-grade, complete, ready-mixed, commercial fertilizer – the best you can get, not the cheapest—and broadcast it on plowed land at the rate of about 800 pounds to the acre ; then harrow it in.

PLANNING AND LAYING-OUT. — Measure the land accurately. Then figure out a definite planting plan, on paper, indicating exactly where everything is to go, and the distance between rows. For convenience draw the plan to scale —say one-sixteenth inch to the foot—and endeavor to have long rows rather than short ones. Therefore, run the rows the *long* way of the garden or field ; whether the rows run north and south or east and west is not so important, although north and south rows are slightly better if they can be conveniently had. Plan to have level, straight rows, rather than elevated little ''beds'' divided by useless paths. Generally speaking, let the spaces between rows be the paths, and keep the entire field as level as possible. (The correct distances apart for rows and plants are given in the chapters on The Strawberry, The Raspberry, etc.)

PLOWING.—Never work soil when it is very wet
and sticky ; wait until it dries into crumbly, workable
condition.   Early spring is the usual time to plow,
although fall-plowing often has advantages when the
ground is soddy or
badly infested with
wireworms, cut-
worms, grubs, etc.
How deep to plow ?
As deep as you
can without bring-
ing up much of the
subsoil.

HOW DEEP TO PLOW ? AS DEEP AS YOU
CAN WITHOUT BRINGING UP MUCH
SUBSOIL

HARROWING.—Follow the plow with the harrow
as soon as possible in the spring.   Do a thorough
job—lengthwise, crosswise and diagonally, until the
ground is as mellow as an ash heap.   The spike-
tooth, spring-tooth, disc and Acme harrows are all
good,—the latter being especially valuable as a
smoothing harrow.

ROLLING OR FLOATING.—After harrowing, it is
often advisable to roll or "float" (smooth) the
ground with some kind of a roller or plank-drag.
This operation crushes lumps and packs down the
soil, but should be followed by a smoothing harrow
that will loosen the surface.   Then the field should
be in excellent condition for marking and planting.

MARKING THE GROUND.—Many growers plant
with a line stretched across the patch and moved into
place for the next row ; this insures absolutely
straight rows, for which I have a great liking, but is
not well adapted to very large fields.   On large

areas, it is very convenient and time-saving to mark
out the entire field in advance of planting.   For
this purpose there are several styles of home-made
markers, one pulled by horse power, and another
kind drawn backward by hand.   By making a few
changes, these markers are easily adjusted to any
width of row desired.

If deep markings are wanted—that is, if a man
wants furrows instead of mere guide marks—a one-

horse plow, or a
cultivator rigged as
a furrower, can be
used by following
the shallow marks
previously made
by a marker.

(Note : In the
Biggle Garden

A HOME-MADE FURROWER

Book several kinds of home-made markers, drags
and floats are illustrated.)

PROPAGATION, PLANTING AND VARIETIES.—In
the chapters on The Strawberry, The Raspberry,
etc., special directions are given covering these
subjects.

TOOLS AND IMPLEMENTS.—Those really needed
are few and not expensive, and they are mentioned
in their appropriate places in this book.   Harriet
hints that the most important things are *thoroughness*
and *persistence*, but I should want to add a good hoe
and cultivator to her list—and a few other
implements.

How many and what kind of tools a fruit grower

will need depends, of course, on the size of his garden. On the very small place the spade or digging fork often takes the place of the plow, the rake doubtless performs the duty of a harrow, elbow grease may be substituted for horse power, and hand hoes, hand cultivators, sprayers, etc., are often

A MACHINE MANURE-SPREADER IS A GREAT HELP
ON LARGE AREAS

substituted for horse-drawn machinery. Fertilizers are applied to square rods or square feet instead of to acres, and manure is perhaps hauled in wheelbarrows and spread by hand instead of in a machine manure-spreader. And the results are as good— sometimes better—than those achieved by the commercial grower with a large acreage.

PLATE II

GANDY

NICK OHMER   SAMPLE

Chapter II

CULTIVATION, MULCHING, IRRIGATION

As for weeds, nip them in the bud.—Tim.

Not only does it take brain work to grow berries successfully, but it requires muscular work as well. But in this, as in most operations of the farm, the brains can save the hands much drudgery.

Any one who does not possess a well-organized brain had better not undertake berry culture, for he will have so much to do with his hands in order to, obtain a compensatory crop, that his efforts will most

CARRY A FILE AND KEEP YOUR HOE SHARP

likely result in failure. He will soon become disgusted and declare that it "does not pay" to grow berries. It is not much bother or work, however, to the one who has a good share of gumption, a little spunk, who was not born tired, and who has a genuine love for the fruit after it is grown.

CULTIVATION.—This is a very important part of the fruit growers' work, and accomplishes the following results :

1. The setting free of plant food by increasing the chemical activities in the soil.

2. The soil is made finer, and hence presents greater surfaces to the roots, thus increasing the area from which the roots can absorb nutriment.

3. The surface of the soil is kept in such condition that it immediately absorbs all the rain

A THOROUGH JOB OF CULTIVATING CAN'T BE DONE BY
GOING BETWEEN ROWS ONLY ONCE. GO TWICE

that falls during the summer, when it is apt to be dry. Little is lost by surface drainage.

4. Moisture is conserved thereby. Where the surface remains undisturbed for weeks the soil becomes packed, so that the moisture from below readily passes to the surface and is evaporated, thus being lost to the growing crop. If the surface is kept light and loose by tillage, so that the capillarity is

broken, but little of the soil moisture comes to the surface and evaporation is not so great. In this way nearly all the moisture remains in the soil, where it can be used by the roots.

5. Thorough tillage has a tendency to cause deeper rooting of the roots. The surface of the soil is made drier by tillage during the early part of the season than it would otherwise be; hence the roots go where the soil is moist. The advantage of deep rooting during drought is obvious.

6. Last but not least, weeds and grass are kept out.

There are a number of excellent horse-cultivators on the market. The Planet Jr. twelve-tooth cultivator and pulverizer is an excellent tool, — especially in the strawberry patch. The teeth are adjustable and those nearest the row may be turned backward, enabling the user to "run shallow" and avoid tearing the roots of the plants. The Iron Age thirteen-tooth cultivator is also very good for such work.

For rough work in very weedy ground, an iron frame, five-tooth cultivator is useful.

A HORSE GRAPE-HOE IS EXCELLENT FOR CLOSE WORK IN A VINEYARD

This implement has several attachments—side shovels, side sweeps, rear hoes, etc.—which are often helpful for special needs; the flat, wide, surface-skimming sweep attachments, I find, are particularly good for killing such weeds as thistles, which are apt to dodge and escape the ordinary narrow cultivator-tooth.

All the cultivators mentioned in the foregoing are adjustable to depth and width, and one horse can pull them easily.

One of the best cultivators I ever had was made with five spring-teeth ; and doubtless there are other good kinds on the market that I have not mentioned.

MULCHING.—There are two kinds of mulch—the dust mulch caused by regular surface cultivation,

and the mulch which is applied in the form of straw, leaves, stable manure, or similar materials. For nearly all purposes I prefer and use the dust mulch. The main object of mulching during the growing season is to prevent the evaporation of moisture in the soil, and shallow cultivation does that effec-

THIS MAN IS APPLYING MULCH SO AS TO HAVE CLEAN STRAWBERRIES

tively ; and does not, like other forms of mulch, furnish breeding places for insects and fungi.

In special instances, however, a mulch of litter is a good thing. For example : Straw, etc., will keep strawberries clean in a fruiting bed ; currant bushes root so near the surface that ordinary cultivation often injures the roots, and therefore a light working of the soil in spring, followed by a heavy mulch of stable manure, is an excellent plan to follow ; in the fall a mulch of strawy manure protects and fertilizes the roots of vines, plants and bushes during cold weather, and prevents the alternate freezing and thawing which causes plants to heave out of the ground more or less.

In later chapters I shall have something more to say about mulching.

IRRIGATION.—Berries are such thirsty plants when loaded with fruit, that ample provision should be made to give them all they can use of water.   In ordinary seasons on most soils this can be done by thorough cultivation or mulching, thus retaining the moisture provided by spring thaws and rains throughout the fruiting season ; but in dry weather the crop is often shortened through lack of water unless irrigation is resorted to.   Unfortunately, however, irrigation is not practicable on the average fruit garden.   But when berries *can* be planted within reach of a stream or pond or well that will yield an abundant supply of water, it will be found advantageous to irrigate ; which will largely increase the crop and greatly lengthen the bearing season.

Hydraulic rams, steam or gasoline engines, windmills, etc., are all used to pump the water where needed ; but if you can run it on the field by gravity, so much the better for you.   It may be applied in furrows opened in the rows (the furrows being filled up as soon as the water has soaked away), or it may

THESE BERRIES WOULD BE LARGER IF THEY HAD BEEN IRRIGATED DURING A DROUGHT

be sprinkled on the plants after sundown with an ordinary hose-nozzle.

The Biggle Orchard Book says on this subject: In regions of normal rainfall, artificial watering is seldom practised or necessary. Cultivation, under normal conditions, conserves sufficient moisture for usual needs. In California and some other states, irrigation is desirable. Folks who need to practise this method should write to the Secretary, U. S. Department of Agriculture, Washington, D. C., and ask for Farmers' Bulletin No. 116, entitled "Irrigation in Fruit Growing."

HINTS I HAVE GATHERED

It pays to carry a file and keep your hoe sharp.

A rope or handle to the harrow helps to guide the implement.

A thorough job of cultivating can't be done by going through a row only once. Cultivate each row at least twice.

It takes an enormous amount of water to irrigate one acre of land; so don't expect to do the job with a small, inexpensive outfit.

For wide rows — such as blackberries, etc.,— it is often possible to use a harrow to better advantage than a cultivator. The Acme does good work here.

In very small gardens a wheel-hoe, to be pushed by hand, has its uses. Then, of course, in

IN VERY SMALL GARDENS A
WHEEL-HOE HAS ITS USES

all kinds of gardens, small or large, the ordinary hand-hoe is indispensable for certain kinds of work.

The important thing is not to allow the weeds to get a *start*. They are easily kept down when young by stirring the soil, but once allowed to gain headway, the labor of fighting them is greatly increased. Weeds kill easiest when the sun shines hot.

I usually set my horse-cultivator to run about two inches deep, for I believe that this plan best conserves moisture and avoids danger of cutting roots. Once in ten days is not too often to cultivate; and always, after a rain, the ground should be stirred with cultivator, rake, harrow or hoe as soon as dry enough to work. (How late in the season to cultivate, and similar special directions, are treated farther on in this book.)

Berries want water; more of it than they are likely to get. Irrigation makes big berries out of what otherwise might be little ones, or helps to make the last picking almost as fine as the first. It makes big, showy berries, but also makes them with less color, soft in texture and not so good in quality as without it. It is more satisfactory to sell water in the berry than in milk, especially after it has been drained from the cow.—J. H. HALE, Connecticut.

Irrigation is beneficial in many ways, but especially so when the fruit is swelling, for berries love moisture, and can not perfect their fruit without it. How and where to apply it has caused many doubts. We have usually let it run between rows on the surface, our land being neither level nor steep. Water runs a long distance without soaking away too soon and without washing. We have never tried plowing a light furrow and laying small underground tiles, but the plan seems feasible for steep side-hills, and not too expensive to be profitable.—J. W. ADAMS, Massachusetts.

PLATE III

BRANDYWINE

MICHEL'S EARLY

CRESCENT

CHAPTER III

SPRAY PUMPS AND FORMULAS

Get after bugs with prayer and a good spray pump.—Tim.

Often I am asked : "What is the best kind of a pump to buy ?" The answer is not an easy one, for very much depends upon a man's needs. For a very small fruit garden, one of the cheap hand-atomizers sold by seedsmen might answer the purpose. These hold about a quart of liquid and cost a dollar or less.

If the garden is of fair size, and yet not too large, one of the compressed-air, shoulder-strap sprayers (several makes are on the market) would be just the thing.

On large areas a more powerful apparatus of greater capacity is required. Perhaps a barrel or tank outfit, mounted on a wagon and worked by hand, would do the job. Or it might be best to in-

A COMPRESSED-AIR SHOULDER-STRAP SPRAYER IS HANDY IN SMALL PATCHES

vest in one of the power outfits that are operated
by either a gasoline engine, geared connection with
wagon wheels, or compressed gas in cylinders.
Some of these machines are designed to spray several
rows of strawberries or two rows of vines at once
automatically.

The ordinary wagon barrel-pump can be used as
a four-row strawberry sprayer, by purchasing a
four-nozzle attachment which can be fastened on to
the back of the wagon and connected with the barrel-
pump. Thus, with a man to pump and a boy to
drive, the work is done thoroughly and automatically
as fast as a team, straddling one row, can walk
along. This attachment costs, complete with nozzles,
connections, etc., but not including pump or barrel,
about $12, and is for sale by several spray-pump
manufacturers. Of course, it is only suited to
large fields.

A good pump should have non-corrosive brass
working parts ; it should be simply made and easily
taken apart for repairs ; it should work easily and
be capable of maintaining a steady, high-pressure
spray from one or several nozzles ; it should be
properly arranged to prevent clogging ; and it should
be provided with some kind of an agitator to keep
the solution in vigorous motion and thoroughly
distributed. A cheap pump is usually a poor invest-
ment. Get a good one.

FUNGICIDES.—Bordeaux mixture is the best and
most useful of all known fungicides for general use.
It is made by taking three pounds of sulphate of
copper, four pounds of quicklime, fifty gallons of

water. First dissolve the copper sulphate. The easiest, quickest way to do this, is to put it into a coarse cloth bag and suspend the bag in a receptacle partly filled with water. Next, slake the lime in a

SPRAYING A YOUNG NINETY-ACRE VINEYARD WITH BORDEAUX, USING A GASOLINE-ENGINE APPARATUS

tub, and strain the milk of lime thus obtained into another receptacle. Now get some one to help you, and, with buckets, *simultaneously* pour the two liquids into the spraying barrel or tank. Lastly, add sufficient water to make fifty gallons.

Powdered sulphur : This is another fungicide ; it is sometimes used for mildew on currant leaves, gooseberry bushes, etc. Dust or blow it on when the plants are wet.

INSECTICIDES.—Of these there are a number of good ones, as follows :

Paris green : Two pounds of quicklime, one-quarter pound of Paris green, fifty gallons of water. The lime helps to neutralize the caustic action of Paris green on tender foliage. Keep mixture well agitated while spraying.

Arsenate of lead : Several ready-prepared, commercial forms of this poison are on the market, and only need dissolving in water ; use about two pounds to fifty gallons of water. For general use I think that arsenate of lead is much better than Paris green ; it sticks better and lasts longer on foliage, remains in suspension in water longer, and it never injures any foliage, even if applied in excessive quantities.

White hellebore : This, if fresh, may be used instead of Paris green or arsenate of lead in some cases—worms on currant and gooseberry bushes, for instance. It is not such a powerful poison as the arsenites, and therefore is safer to use in the family garden. Steep two ounces in one gallon of hot water, and use as a spray, stirring it often.

BORDEAUX COMBINED WITH INSECT POISON. — By adding one-quarter pound of Paris green to each fifty gallons of the Bordeaux formula, the mixture becomes a combined fungicide *and* insecticide. Or, instead of Paris green, add about two pounds of arsenate of lead. With this combination the fruit grower is able to fight fungous diseases and most insect pests with *one* spray, and I trust that my readers will often avail themselves of this useful team; if carefully handled, they'll pull nicely together and do their work well.

FOR SUCKING INSECTS. — Now we come to another class of insecticides, suited to insects which suck a plant's juices but do not chew. Arsenic will not kill such pests; therefore we must resort to solutions which kill by *contact*. Here are some of the best-known recipes of this kind :

Kerosene emulsion : One-half pound of hard or one quart of soft soap ; kerosene, two gallons ; boiling soft water, one gallon. If hard soap is used, slice it fine and dissolve in water by boiling ; add the boiling solution (away from the fire) to the kerosene, and stir or violently churn for from five to eight minutes, until the mixture assumes a creamy consistency. If a spray pump is at hand, pump the mixture back upon itself with considerable force for about five minutes. Keep this as a stock. *It must be further diluted with soft water before using.* One part of emulsion to fifteen parts of water is about right for lice.

Whale-oil soap solution : Dissolve one pound of whale-oil soap in a gallon of hot water, and dilute with about six gallons of cold water. This is a good application for aphis (lice) and scale insects.

Tobacco tea : This solution may be prepared by placing five pounds of tobacco stems in a water-tight vessel, and then covering them with three gallons of hot water. Allow to stand several hours ; dilute the

TIMELY SPRAYING SAVED THESE BERRIES FROM BUGS AND BLIGHT

liquor by adding about seven gallons of water. Strain and apply. Good for lice.

Pyrethrum : This is also known as buhach, or Persian insect powder. The best is called California buhach ; the imported powder is not so fresh as a rule, and therefore not so strong. It may be used as a dry powder, dusted on with a powder bellows when the plants are wet ; or one ounce of it may be steeped in one gallon of water, and sprayed on the plants or vines at any time. A good lice remedy.

Special Note : The various insect and fungous pests which attack small fruits are specifically mentioned in the chapters on The Strawberry, The Grape, etc., together with the best remedies to use in each case.

SPRAYING NOTES

After spraying, pump water through the pump and hose to clean them of the mixture, so that it shall not needlessly corrode them.

All spraying mixtures should be constantly agitated when in use. If this is not done, some of the ingredients, particularly Paris green, is apt to settle to the bottom of tank or barrel.

Remember that fungicides are not cures, but *preventives*. It is important to begin their use early in the season before the trouble begins, and repeat the application several times at intervals.

Bees are excellent friends of the fruit grower, because they help to pollinate berry blossoms. Therefore this rule is a good one : Never spray fruit vines or bushes when they are in blossom, for fear of killing the bees.

All spraying mixtures should be strained before using, to prevent clogging the nozzles. A box, with the bottom and top knocked out, will make a frame for a strainer; a brass-wire mesh—eighteen or twenty meshes to the inch—can be securely tacked around the bottom to complete the job.

It must be remembered that most spraying materials are poisonous and should be so labeled. If ordinary precautions are taken, there is no danger attending their application. Properly-sprayed fruit (that is, fruit not sprayed too near the time of maturity) is, on account of the great dilution and the action of rains, perfectly safe to eat.

The Vermorel nozzle is very popular; so are several other makes that I have tried. The main thing to demand is a nozzle

A GOOD NOZZLE SHOULD THROW A
FINE MIST, LIKE STEAM

that will throw a fine mist, like steam, which settles on the plants like dew. A sprinkler, resulting in much drip upon the ground, is not wanted. And, too, a good nozzle should not clog easily, and when it does clog it should be quickly cleanable.

The lime-sulphur mixture is the standard remedy for the San Jose scale. Seedsmen sell it by the quart or gallon. The time to use it is *after* the leaves are off —in the late fall or early spring. Currant bushes, etc., if attacked by this pest, should have prompt treatment. It is a round, dark scale with a central dot or nipple, and is not easy to see without the aid of a magnifying glass. (Note: Seedsmen also sell ready-prepared Bordeaux, kerosene emulsion, Bordeaux-arsenate of lead, etc. So it isn't necessary to make these sprays at home unless you prefer to do so.)

PLATE IV

CLYDE

LADY THOMPSON

ENORMOUS

Chapter IV

THE STRAWBERRY

" Doubtless the Lord might have made a better fruit than the strawberry, but doubtless He never did."

Being the first fruit to ripen, the strawberry comes to the table as a welcome visitor when the appetite is capricious. So beautiful in form, color and fragrance, it is among fruits what the rose is among flowers. In flavor so delicious, in healthfulness so beneficial, that invalids often gain strength while its season lasts. Strawberries fully ripe and freshly picked from the vines may be eaten at every meal, in saucers heaped high like pyramids, and will usually nourish the most delicate stomachs.

The charms of the strawberry do not all end in the eating of it. No fruit is so soon produced after being planted. It affords employment—pleasant, easy and profitable—for poor men with little land ; for old men with little physical strength ; for women, boys and girls who love to till the soil and delve in mother earth.

PROFIT IN STRAWBERRIES.—Novices in berry culture will be surprised to know that more bushels of strawberries can be grown on an acre than of wheat or corn, but such is the fact, as testified to by many experienced growers.

Prof. Bailey, in his "Horticulturist's Rule Book," says that the average yield of an acre of strawberries is from seventy-five to 300 bushels.

Mr. Rosa, Delaware, reports that from one and one-fourth acres he sold, one summer, 169 crates of thirty-two quarts each, or 5,408 quarts. They were sold in Philadelphia and paid him net, not counting the picking, $653.79, or more than twelve cents per quart.

PICKING MARSHALLS FOR
MOTHER'S DESSERT

From a patch only 108 x 213 feet in size, W. E. Pennypacker, Pennsylvania, recently sold in one season, 4,721 quarts for $461.

From Boise, Idaho, comes an interesting strawberry item, relative to the patch of J. H. Waite, located a short distance from that city. Mr. Waite's patch measured just a little short of an acre and a quarter. He marketed his first berries June 1st, and from that date until July 7th he brought in 12,798 boxes, from which he realized $807.70.

And here's a big report from California : "We began to market our strawberries on March 30, 1907, and picked every day until October 25th. The area was two and three-quarter acres, and the variety Brandywine. The sales were 79,000 baskets (pints), for which we received $5,000. Such yields are not possible in the East, however. I am a Pennsylvanian, and know."—Q. A. LOBINGIER.

I myself have grown strawberries at the rate of 200 bushels per acre ; but one year I expected 300 bushels and got about fifty. It is never safe to count on too much, nor to be too sure of results, nor to increase the acreage unduly. Small patches, as a rule, are much more profitable than large ones.

The expense of bringing an acre of berries into profitable bearing is greater than most folks think. There's the interest or rent on the land, the value of the plants set (whether you buy or raise them), the cost of heavy fertilizing, horse hire or keep, labor, wear and tear on implements, etc. "You may safely estimate," says a successful Wisconsin grower, "that every acre of good small fruits, well set, missing hills filled in and brought to a bearing age, will cost from $100 to $150, or an equivalent in honest work at $1.50 per day."

THIS MAN HAS A STRAWBERRY
PATCH RIGHT AT HIS DOOR

SOIL AND LOCATION. — The strawberry will adapt itself to a great variety of soils and location. It is grown successfully in every state in the Union, as it is prized by the people everywhere. Different varieties require somewhat different conditions in climate and soil; thus one that thrives on sandy land may not do so well on clay; and certain kinds that succeed in northern latitudes will not stand a hot southern sun ; and vice versa.

Almost any soil that will produce a good crop of corn or potatoes will give fair returns with strawberries ; land inclined to be moist (but not too wet) and not subject to injury by drought, will be best.

If early bearing is wanted, take an early variety, set on sunny southwest-lying land. If you want late fruit, take a late variety, set on an east or northeast slope and allow the mulch to remain as long as possible.

Be sure to read, carefully, the various hints given in Chapter I about soils, altitude, drainage, fertilizing, plowing, harrowing, marking, etc., and, especially, pay heed to the remarks about ''preliminary preparation '' and the importance of not setting berries on land which has recently been in grass sod.

Generally speaking, a well-drained clay-loam, filled with humus, is an ideal strawberry soil ; but this fruit often does well on more sandy soils. A heavy, stiff clay is least desirable, in my opinion.

For raising plants to sell, says A. I. Root, Ohio, I should prefer low bottom land inclining to sand, made very rich with manure ; but for raising berries I should take upland, turn under clover sod and work in all the stable manure I could get hold of. There is practically no such thing as making it too rich.

OBTAINING PLANTS TO SET.—To grow strawberries successfully, beginners should order their plants of some reliable nurseryman very early in the spring. If a dozen, thirty or a hundred plants only are wanted, they can be sent by mail. Five hundred, or more, should go by express. If ordered early the nurseryman will send them as soon as the ground is fit for planting. (The following year the off-shoots or runners from these first plants should furnish a surplus of plants for setting a second bed, and so on, indefinitely.)

When plants are received by mail or express from a distance, they should be opened at once and the roots should be dipped in water. If the ground is not ready for them, break open the bunches, spread out the roots, and "heel them in" closely together in moist earth in a sheltered spot outdoors, or in the cellar. Heeling-in means simply a temporary plant-

A REFRESHING TUMBLERFUL
(Cluster of Gandys)

ing, plants touching each other, and placed in a half-reclining position. In this way, if watered and shaded, they may be safely kept several days or weeks.

If the plants are in plant beds of your own or a neighbor's raising, dig up the whole row, throwing out the old plants. If plants must be taken from a fruiting bed, you can dig from the side of the rows. But, remember, this last method means that you'll get only the smaller, weaker plants ; it is much better to dig an entire row as far as necessary, even in a fruiting bed. Why plant ''little potatoes'' when you can get larger ones?

A potato hook is a good tool with which to dig strawberry plants. As fast as shaken from the soil gather them up. Hold the plants in the left hand, crowns of the plants as nearly even as possible, and when the hand is full trim off all runners and dead or diseased leaves, and lay the plants in a basket, roots straight and all one way.

Take the plants to the packing house. Tie them in bunches and dip in water, and if to be shipped, pack in moss and forward as soon as possible. If to be set out at home, the tip ends of the roots are cut or sheared off before setting; it is customary to remove, in this way, the lower one-quarter or one-third of all strawberry roots,— as an aid to new root formation. Some growers like to heel-in all plants a day or two before setting, claiming that this preliminary treatment puts the plants in better condition to stand the setting ordeal; I'm inclined to think that they are right, although most of us are usually in too big a hurry to wait.

Caution : As a general thing it is better not to set plants from an old bed which has borne even one crop of fruit. Plants from such beds are often not so full of vigor and health as those from a patch which has never produced berries. Also, from any bed, see to it that all little, feeble plants and all old or ''parent'' plants are thrown out. The dark color of the roots is a distinguishing mark of old plants. Such plants are worthless, and if any are discovered

PLATE V.

MARSHALL

AROMA   WILLIAM BELT

in packages sent from a nursery, they should be
thrown away ; it is useless to set them.

Southern people who wish to buy northern-grown
plants should do so late in the fall. They can not
get them early enough in the spring, and their summer and early fall are too hot for setting plants
grown in the North.

STAMINATES AND PISTILLATES. —Beginners may
need to be told that the staminate plants are those
which have both stamens and pistils and which carry
their own pollen ; they are, therefore, called perfect
flowering or bisexual. Blossoms of real pistillates

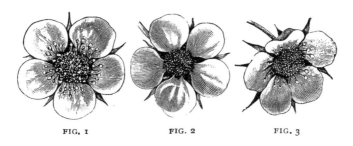

FIG. 1   FIG. 2   FIG. 3

contain no pollen, are imperfect flowering because
lacking in stamens, and, therefore, require the aid of
a staminate variety before they will produce fruit.
A strong staminate blossom is shown in Fig. 1, a
pistillate in Fig. 2, while a feeble staminate is indicated in Fig. 3, which has a few undeveloped stamens
only. Staminates can be grown in a bed by themselves, and will bear fruit ; real pistillates are fruitless,
unless they have staminates nearby to pollinate them.
The necessary pollen is carried from staminates to

pistillates by the aid of the wind and of bees ; rainy weather in blossoming time is apt to interfere with the distribution of pollen, and cause an imperfect crop of fruit, in which many specimens are shortened at the apex, and made small and ill-formed. Wet weather likewise interrupts the perfect development of fruit on staminate varieties, but to a less extent than on pistillates.

It is a question often discussed among berry growers, whether it is not best to discard the imperfect flowering varieties entirely, owing to the inconvenience of having to plant a suitable pollenizer near them ; but most growers have found that pistillates pay, because they produce *more fruit*—when properly pollinated—than the staminates. I have found this so, myself, and always set my bed in this way : One row of staminates, two rows of pistillates, then one row of staminates,—and repeat this order throughout the patch. Thus each double row of imperfect flowers has a row of perfect flowers on each side of it. Some growers, however, prefer to set the pistillates and staminates in alternate rows.

I would call attention to the following facts : Early spring frosts are more apt to injure the staminate blossoms than the pistillate. Some varieties, notably Haverland, which is considered a pistillate, have a *little* pollen of their own, and require less care in planting a staminate variety near them ; in fact, the Haverland will almost fertilize itself. There are other so-called pistillates with similar capacity, especially in favorable seasons.

Care must be taken that the pollenizer be a sort

that will bloom abundantly, and at the right time, so that the adjacent pistillate blossoms may receive pollen throughout the blossoming period. For this some varieties of staminates are much better than others, and some are quite inadequate. It is important, also, that the staminates and pistillates to go together should be selected so that the fruit will ripen at about the same time. For instance, the Parker Earle is well adapted to fertilize the Haverland, being of the same form and ripening nearly at the same time. It is probable that every desirable pistillate sort has a good friend among the staminates that it should be married to in preference to the others, and the wide-awake berryman will look sharp that his varieties be well mated.

A pistillate variety will vary quite perceptibly when fertilized by different perfect varieties ; so, if you want firmness, you should fertilize with a firm berry ; if sweetness is wanted, fertilize with a sweet one ; if dark color is wanted, fertilize with a dark one. In fact, whatever peculiarity you wish to transmit to the pistillate variety, seek it in the perfect variety you would fertilize with. Staminates affect the size, color, solidity, shape and quality of pistillates. Make a study of which varieties planted together bring the best results.

The honey-bee will visit 10,000 strawberry blossoms in a single day, and thus does valuable cross-fertilization work for the berry grower.

DISTANCE APART.—The right distance apart of rows and plants depends upon the method of growing—that is, which system of growing you choose.

Some of the systems are here illustrated, but are treated more fully in the next chapter.

To ascertain how many plants are required for an acre, multiply the distance apart of the rows in feet by the distance apart of the plants in the rows, and

HILL SYSTEM
(Just started)

HILL SYSTEM
(Six months' growth)

NARROW MATTED-ROW
SYSTEM

WIDE MATTED-ROWS
(Not a good system)

divide the product into 43,560. Thus, if the rows are four feet apart and the plants two feet, it will take 5,445 to plant an acre.

Special note : Some varieties produce more runners than others, and certain kinds have longer runners than others ; therefore the exact distance apart to set plants in the row depends somewhat on the variety as well as upon the system.

TIME TO SET PLANTS.—J. H. Hale, Connecticut, and A. I. Root, Ohio, both write me that early spring is the best time. As soon as the ground can be worked – the earlier the better – is my own rule; which means early April in Pennsylvania. A friend writes from Missouri that March 1st is the best time to set strawberries there.

In the South, planting can be done still earlier—depending, of course, upon the exact degree of lati-

THE HONEY-BEE VISITS THOUSANDS OF BLOSSOMS IN A SINGLE DAY, AND THUS DOES VALUABLE CROSS-FERTILIZATION WORK.

tude. February is a favorite month in some southern states, although late fall planting is practised by many southern growers. Near the Gulf, plants are often set in late summer during rainy weather.

Now we come to the question of late-summer or early-fall set plants in the North. Some growers claim that by the use of pot-grown plants set, say, in August, they can save one season's time and yet get good crops of big berries the following June.

Yes, this is possible ; it has been done, it may be done again.　But pot-grown plants are expensive, and the labor of setting many plants closely together (for no runners can be expected to help fill out rows) is excessive.　Personally, I prefer spring setting with

METHOD OF OBTAINING
POTTED PLANTS

ordinary plants, and thousands of growers will agree with me in this.　Those who wish to try the fall planting method, however, will find additional information about pot-grown plants in the next chapter.

PLATE VI.

BEDERWOOD

LOVETT

CLIMAX

CHAPTER V

THE STRAWBERRY (CONTINUED)

Halve your acreage and double your fertilizer.—Tim.

Presuming that you have the plants ready, the ground prepared and marked as suggested in Chapter I, I will now proceed to the operation of setting, — and an important one it is, too.

There are at least five good ways to grow strawberries, viz., the hill system, the single, double and

NICE BERRIES, BUT THE BOX IS "SLACK PACKED"

triple hedge-row systems, and the narrow matted-row system. It is not the quantity, numerically speaking, but the quality and size of the berries that count ; and to produce large well-colored and highly-flavored berries the plants must have ample room to develop and admit of cultivation.

The method of growing strawberries in wide, thickly-matted rows, says Geo. W. Stephens, Iowa, or where the vines are allowed to spread all over the ground, can not be recommended, because no cultivation is possible in the row, and consequently the ground soon dries out, and the result is a lot of crowded and stunted plants that will yield less than half a crop of small berries, and the second year hardly any. (See picture of wide matted-rows in Chapter IV.) This is the reason so many growers set out a new patch every year or two and plow up the old one. When grown by the hill or hedge-row systems the plants may bear good crops for from three to five years or even longer, and, where the fertility of the soil is kept up, many crowns will be capable of yielding enormous crops of berries. Some varieties do their best only in hills.

For a hedge-row, mark out the rows two and one-half feet apart and set the plants about two feet in the

SINGLE HEDGE-ROW

row. (This is for horse cultivation ; two feet is ample for hand-hoe or wheel-hoe work.) Then allow two runners to grow from each plant and layer them in the row in a straight line with the mother plants. When the row is completed the plants will be about eight inches apart in the row. (See illustration.)

For a double hedge-row place the rows three feet apart and form an ordinary hedge-row. Then allow one runner to grow from each plant, layering them along one side, forming a second hedge-row eight or nine inches from the first, as indicated in cut.

DOUBLE HEDGE-ROW

For a triple hedge-row place the rows three and one-half feet apart and the plants about two feet in the row. (Rows may be six inches narrower for hoe work.) Allow four runners to grow from the mother plant, layering two of them in the row and one at each side.

TRIPLE HEDGE-ROW

Then let each of the four runner plants throw out one runner, and layer them in the two outside rows as shown in the drawing.

By the last two methods, no runner-plant is allowed to exhaust itself by throwing out more than one runner, and all will be earlier and stronger than they otherwise would. Layer the runners in regular order so that they will be straight in the row and in line across the row.

Narrow matted-row system : This is practically the same as the triple hedge-row, except that little or no care is used in placing the runners,—which are allowed to form and set almost at will until a compact row about eighteen inches wide results ; then any runners which straggle outside the row are cut off. (See illustration in Chapter IV.) This system, being less trouble than a systematic hedge-row, is popular with many growers ; and if such a bed is kept only one or two years, it usually proves profitable. Its weak point lies in the fact that too many plants are apt to set in the row unless you are careful, and crowding means smaller fruit. Rightly handled— and keeping such a bed only one year – I have found the narrow matted-row a satisfactory system on my ground.

Hill system of culture : Set the plants about fifteen inches apart in rows about three feet apart, for horse cultivation, or in two-foot rows for hoe work only. *All* runners are promptly and regularly cut off all through the growing season, and the plants, not being exhausted by runner bearing, put all their strength into themselves and grow big, sturdy and bushy. (See two illustrations in Chapter IV.) This method requires high culture and fertilization to produce satisfactory results. In addition to other fertilizing, occasional applications of nitrate of soda the first season, and again the following spring after growth starts, are helpful. Nitrate of soda is useful in any system of strawberry growing, but must be used with care ; it is a powerful stimulant and too much may harm the plants. It should be applied in small doses often. About 100 pounds at a dose to the acre is usually enough, and, generally speaking, it should not come in direct contact with plants. One pound is enough for about 100 feet of row.

Another excellent way of growing plants in hills, is to set them 20 x 20 inches apart each way, and leave a thirty-inch aisle between each three rows.

SETTING PLANTS. —Have a boy to carry the plants, roots down, in a pail with a little water in it. Have him drop or place the plants as fast as needed – and no faster. Openings for the plants can be made at proper distances along the row with an ordinary spade. Force it into the soil, upright, push it slightly from you, and the opening is made. Into this put the roots, spreading them out fan-shape. Be careful not to set too deep as in Fig. 1, or too shallow as in Fig. 2,

and do not bunch the roots as in Fig. 3, but see
that every one goes in like Fig. 4. Press the soil
firmly around the roots and tread it down with the

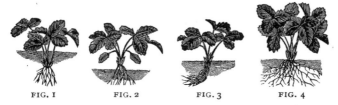

FIG. 1   FIG. 2   FIG. 3   FIG. 4

feet — *this is important.* Also nip off any runners,
blossoms or unhealthy leaves that may not have been
previously removed. A special trowel, Fig. 5, flat
like a mason's trowel, but wide and full at the point,
with extra long handle, is a tool used by some growers for setting. Some other growers use dibbers.

If the plants are in good condition, the soil moist,
and the setting properly done at the right season,
they should not need shading. Do the setting
toward evening, if you can ; then they have all night
to recuperate in before the heat of another
day. In a small garden, however, it is
often a help to shade the newly-set plants
for a few days ; the small grower can
utilize shingles, newspapers, berry boxes,
etc., etc., for this worthy purpose.

Two men and a boy, working together,   FIG. 5
can do fast work setting plants in large fields. The
first man makes the holes with a spade or other tool,
the boy drops the plants, and the second man, on his
knees, places the plants in position and pulls the soil
around the roots. The first man, after making a few

holes ahead of the planter, can go back occasionally and firm the soil around the plants with his feet— returning again to his hole-making job.   Transplanting machines drawn by horses are sometimes used for very large areas, and these are supplied with a watering device.

CULTIVATION.—As soon as the field is planted, start the cultivator or hoe at once, so as to loosen up the top two inches of the trodden land and stop the

A STRAWBERRY-SETTING SCENE IN NEW JERSEY

evaporation of soil moisture.   Cultivation should continue at regular intervals, say ten days apart, until the ground freezes in the fall.   As the rows widen, narrow down the cultivator and depend upon hand-hoeing and hand-pulling to keep out weeds inside the row.

Several kinds of claw-like hand weeders are on the market, and for close work around plants are a great saving of skin and nails.   Harriet says that steel is cheaper than finger tips.   Various shaped hoes are favored by different growers, but for myself a hoe like Fig. 6 is good enough for general work ; the blade is only three inches wide. However, for work inside of full rows, I should want —in addition—an extra hoe cut with a file into the shape shown by the dotted lines in Fig. 7. This makes a very excellent strawberry hoe, with various cutting edges.

FIG. 6

In late autumn make surface ditches or furrows to catch the winter's surface waters and so prevent water standing on the plants. Of course, when the ground is frozen, water can not soak into tile drains even if you have them. A little study will show where to run the ditches so as most surely to lead away

FIG. 7

overflows. If not needed longer in spring they can be closed up. They need generally be no bigger than furrows. Tile underdrains are often needed in the berry patch, also, and the two supplement each other very nicely.

BLOSSOMS AND FRUIT.—Allow no blossoms or fruit the first season. Premature bearing weakens the plants, and to permit it is poor economy.

CUTTING RUNNERS. — A *sharp* hoe is the only runner cutter used on my patch. Various patented contrivances have been exploited at different times, but the hoe is sufficiently handy and efficient for most people. For plants grown in hills, it might pay to go to the expense of a specially-made runner cutter. A round hoop of steel, thin and sharp at the bottom, and connected with an upright long handle, is a tool which any blacksmith could make to your order. It should be large enough to fit around a plant with-

A WELL-SET, WELL-CARED-FOR BACKYARD STRAWBERRY PATCH IN AN OHIO TOWN

out harming the leaves ; then, by pressing down on
the handle, the sharpened bottom will cut off runners
on all sides at one clip, - on the principle of the tin
cooky-cutter which Harriet uses in the kitchen.

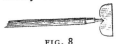

Some folks make a serviceable
runner cutter by bending a hoe
into the shape shown in Fig. 8.

FIG. 8

While I'm on this subject of runner cutting, let
me say that if you are growing a hedge or matted row,
do not make the mistake of cutting off the first set of
runners that the plants throw out.   Root these *first*
runners, for the sooner runners can be rooted, the
longer they will have to grow and perfect fruit buds
for next season's crop.   A while ago, many growers
advocated cutting the early runners so as to strengthen
the plants, and then rooting the *second* lot of runners.

Experience has shown that this plan is usually a
mistake.   *Root the early runners and cut off the later
ones after the row is formed.*

POTTED PLANTS FOR LATE SETTING.—For late-
summer or early-fall planting, which is usually a dry
time, it is very important that young plants for setting
should be removed without cutting or even disturbing
the roots.   Small pots are used into which the roots
are induced to grow.   They must not, however, be
allowed to remain until they are too compactly
rooted,—that is, pot-bound.   August is generally the
best time to set pot-grown plants.

A woodcut shown in Chapter IV represents the
method of potting runners.   Pot-grown plants can be
had of most nurserymen ; or you can easily raise
them yourself by buying a number of tiny pots, filling

them with earth, sinking them alongside a new strawberry bed, and causing a runner to take root in each pot. Of course this potting must be done a number of weeks before setting time. To facilitate the rooting of a runner in a pot, press the runner slightly into the soil in centre, and then hold it in place with a small stone. It will soon root and fill the pot with roots ; then, when ready, the runner which holds it to the parent plant should be severed and the pot lifted from the soil and removed to the setting field. There the plants should be watered before setting.

For this late setting begin to prepare the ground a few weeks ahead of planting time. Plow it early so that it will have a chance to settle. Harrow or rake it often, to keep down weeds and conserve moisture. Then the bed will be in fine shape to receive the plants.

Space the plants as advised under ''hill system of culture.'' When setting a plant, simply invert the pot, jar the plant loose, and carefully remove it from the pot ; then without disturbing the roots, set the plant in the hole where it is to go and press soil firmly around it,—being careful not to cover the crown.

Keep off all runners, if any form ; cultivate until the ground freezes ; and then mulch the bed. A fair crop of berries may be expected the following June.

As already stated I do not advocate August or September setting in the North as a general thing. But in a small way in the home garden, or for the man who wasn't able to set plants in the spring, or for those who want to get a quick test of some new variety, the plan is sometimes worth following.

In closing this subject, permit me to say that *if* all conditions are favorable (which they seldom are), it is possible to set strawberry plants in August without a preliminary potting treatment.   In this connection a friend writes :   '' In our own garden, our land being somewhat inclined to clay, we can take up the plants with a round trowel with a lump of soil adhering and thus remove them to their new quarters without loss.

SHADING NEWLY-SET PLANTS
WITH BERRY BOXES

Their growth will not in the least be retarded.   The best crop we have ever produced was from plants set out on the nineteenth day of August, the plat being 250 feet long and five rows wide.   When planted in the spring it requires the best part of two seasons to perfect a large yield, thus losing the use of the land for one entire season and adding much to the labor for so much longer a period ; for the cost of cultivating so short a time in hills is trifling compared with hoeing and weeding where runners are permitted to grow.''

HOTHOUSE BERRIES.—Few commercial growers attempt to force winter berries under glass, because southern-grown strawberries now reach northern markets early and cheaply.   Hothouse berries must be pollinated by hand and it is troublesome and expensive to grow them successfully.   Pot-grown plants, allowed to freeze outdoors, and later moved indoors, are best to use. The temperature should not be very warm at first ; increase the heat by degrees.

MULCHING.—The importance of mulching is becoming better understood than formerly, and the work is done with more thoroughness. Early winter is the best time to do the work, after the ground becomes hard enough to bear a team. Swamp hay, straw and cut corn-fodder are all good materials for the purpose. The plants should be covered up out of sight during the winter, and in the spring the mulch should be " loosened up " and only part of it allowed to remain. The surplus can be forked into the aisles between rows.

Late berries : A heavy mulch left on extra late in the spring insures late berries (if you want them). The plants must have some vent, if covered deeply, after the weather warms up, but do not rake the mulch off the row. For a late crop of berries four inches is not too deep for the mulch.

Taking the mulch off too soon is a fruitful cause of injury from frost.

SPRING CULTIVATION.—This is a delusion and a snare. Keep the ground of a fruiting bed moist and mellow by a suitable mulch, not by cultivation ; and pull out by hand, if you wish, any weeds which succeed in pushing through the mulch.

PLATE VII.

BUBACH

EXCELSIOR

CHAPTER VI

THE STRAWBERRY (CONCLUDED)

The family is entitled to a daily feast of strawberries in season.—Harriet.

Now I'll tell you how to renovate old beds. By old beds I mean any strawberry bed which has just produced a crop of berries. If you run your patch according to the hill system or any of the hedge-row systems already mentioned, you can perhaps keep the bed several years to advantage. Proceed thus: The second year and each year thereafter as long as

TAKE ONE!

the plants are fruited, as soon as the picking is over remove the mulch and mow the leaves and rake them into the aisles and burn them, or burn mulch and leaves as advised in the next paragraph ; then cultivate the same as the first year, keeping off all runners the entire season. Apply winter mulch as before.

For the matted-row system, the following plan is often recommended : After fruiting time mow off the plant leaves and tops, let them dry a day or so, rake them into the aisles, and then set fire to dry leaves and mulch. Choose a dry, windy day for this job, so that the fire will run quickly along the rows. (Note : Some growers advocate burning the leaves and mulch directly *on* the fruiting rows ; but experience has shown that the plants are thus likely to be damaged unless all conditions are just right. It's too big a risk, I think.) As soon as the new growth starts, narrow down the rows by plowing one furrow away from each side, and then cultivate the soil back into place. Then treat the bed the same as the first year. When it has borne two crops, better plow it under and set the field to sweet corn or something else. The finest berries generally grow on new beds. Instead of plowing away from the rows, some growers prefer to turn furrows from the aisles *over upon* the rows, and then harrow the field lengthwise and cross-wise sufficiently to uncover the plants and partially level the ground.

Generally speaking, it is better not to follow strawberries with strawberries, if you can help it, in the same spot until the soil has been "rested" or rotated a year or so with other crops. Remember that an old bed is a prolific breeding place for strawberry fungi and insects. A new bed each year in a new place is usually the safest policy ; but there are exceptions to this rule.

INSECTS AND DISEASES. — Leaf-spot, blight, rust and mildew can all be largely controlled by early sprayings with Bordeaux mixture. Rotation of crops and the annual mowing and burning of each bed after it has fruited, will usually control the crown-borer, the leaf-roller, and similar insects ; also, put some arsenate of lead in the Bordeaux mixture and use the combined spray until little green berries begin to form, then stop,

NO BLIGHT OR GRUBS IN THIS PATCH

and resume spraying after the fruiting season is over. White grubs (larvæ of May beetles, or "June bugs," as they are sometimes called) often attack the roots below ground, and the plant withers and dies ; there are no good remedies, but fall-plowing and regular cultivation are preventives ; never plant strawberries on ground which has been in sod within two or three years and you'll have little trouble with white grubs.

Some varieties are more liable to rust, etc., than others, and the trouble appears to be greater in some neighborhoods than in others. If possible, plant varieties which are least subject to fungous troubles. Keeping a bed only one season, rotation in planting, and using only strong, healthy plants for setting, are all helpful in fighting these diseases. Remember, Bordeaux should be applied early in the spring, and again after the blossoms fall. For the newly-set bed apply as often as there is any sign of rust.

Root lice often appear in great numbers, feeding

on the roots of the plants. Plants received from nurseries should always be examined, and if lice are on them, they should be dipped in kerosene emulsion.

It is best to be watchful of all destructive insects, and where any of them are very troublesome, change plants and ground, burn the bed over after fruiting and plow down.

OTHER PESTS.—Moles in their search for white grubs, often burrow along a row of plants and damage them by heaving them up. One or two mole traps, for sale by seedsmen and implement dealers, will prevent this trouble if it becomes serious. The moles, however, are doing you a good turn in one way—they eat the grubs.

Robins and other birds take their toll from the berry patch, but on large fields the proportion of loss caused by them is so small that commercial growers disregard it,—remembering the great good these same birds do in the way of destroying insects. On a small patch, though, the damage from birds is sometimes serious enough to warrant special measures. Cheap mosquito-netting might be spread over a small patch ; or around currant bushes. I know of one strawberry lover who screens in quite a fair-sized bed of strawberries ; he uses wire poultry-netting, supported, top and sides, six feet high, and leaves the netting there permanently. Another grower uses cotton netting placed along the rows and removes it when the pickers are at work.

Hens are a pest in the berry field at fruiting time; keep them out.

Toads are friends of the gardener, because of the

many insects they devour. Never kill or drive away a toad. Lady-bugs, or lady-birds, are friends, too; they eat the tiny lice that suck a plant's juices. Bees are beneficial, for they help the pollination of blossoms.

Frost injury : A heavy frost when the vines are in blossom is often a serious thing to the grower. Smudge fires — and the resulting smoke – may save a berry patch. Some growers mix coal-tar with saw-dust and old straw, and put the mixture in heaps around the patch, in readiness for an emergency. Other growers burn crude oil in iron pots sold for that purpose. Sprinkling or irrigating a patch when frost threatens, will sometimes save it. Or it may be saved by forking the straw mulch from the aisles over on to the vines, —leaving it there for a few hours or a day or two.

VARIETIES.—Some growers make no distinction between the varieties intended for market purposes and those for the family to use ; nevertheless, I believe that different sorts for each purpose might advantageously be selected, because it is not always the sweetest berry that will yield the greatest number of quarts, nor carry to market in the most salable condition. On the other hand, the variety that is most desirable for the family to feast on may be a light yielder, and perhaps of poor color and soft in texture. Buyers in the towns are attracted by size, color and freshness, and are not very particular about the flavor; while for the folks at home nothing is too good for them. A large number of varieties that have high merit as home fruit will not carry to market in good order, and should not be placed in the market list. However, there are some varieties which are well suited for both home and market.

"The strawberry plant," says a veteran grower, "indicates by its leaf what is the shade of color, size, shape and quality of the berry. The lighter the color of the leaf, the lighter you will find the color of the berry; the darker the leaf, the darker the berry. The leaf also indicates the size of the berry. An irregular berry is indicated by an irregular leaf, a round berry by a round leaf, a long berry by a long leaf. Leaves on the same plant will vary considerably, no two are alike, but their general form will be the same. Also the relative productiveness of different varieties of strawberries can be told by the number of serratures, or saw teeth, on the leaf. The greater the number of serratures the greater the number of berries that will be produced on an individual plant."

Do varieties "run out"? For my own part I do not believe that varieties will run out if proper intelligence is given to their propagation by runners and to their after culture. Carelessness and ignorance on the part of the growers, and enterprise in those who have made it their duty to introduce new varieties, are the main causes of strawberry deterioration ; or, they run out because plant nurserymen cease to sell them, and because improvement is the order of the day, and new and better kinds are discovered or propagated. One of the greatest errors made by strawberry growers is the discarding of valuable kinds before they give them a fair trial and learn just what treatment is best for them, to take up with some new and costly variety, which, in due time, will go out in the same manner, perhaps being inferior in every way to the old sorts.   In this way many have already discarded that wonderful berry, the Gandy, which succeeds admirably where brains are applied to its culture, and the required conditions of a crop are complied with.

HOW DO YOU LIKE THESE?   ONLY TWELVE IN A QUART BOX

Breeding up varieties :   Plants, like animals, may be bred up to a higher efficiency by a careful and systematic selection of parents.   By always taking for setting purposes the best and strongest plants, you can in a few years improve the general vigor of almost any variety.   The same thing is done with corn, or potatoes, or other things.

Producing new kinds :   As a rule, new varieties are accidents or ''sports'' found among seedling plants. A few enthusiastic horticulturists grow experimental beds of plants from strawberry seeds each year, hoping to find a seedling which has superior merit. Generally their hopes are not rewarded, for most seedlings are inferior to their parents ; but, once in a great while, a lucky find is made and a new variety appears on the market. Expert horticulturists sometimes succeed in producing new kinds of more or less merit, by artificially cross-fertilizing the blossoms of two selected varieties of opposite sex kept by themselves (using a camel's-hair brush to transfer the pollen), and later planting the resulting seeds from the female or pistillate variety and choosing the most promising plants therefrom. Seeds should be thoroughly dried and cured, and may be planted in the fall.

"What variety shall I plant?" Hundreds of growers ask this question every year, and, unfortunately, no correct general answer is possible ; for each locality, soil or climate is a law unto itself. A variety which does well in one place, may not do so well in another ; and, also, some localities and markets prefer certain kinds of berries. Therefore, to answer the foregoing question in the most helpful manner, I have written to strawberry growers in various states and asked them to give a brief list of the best kinds for general use in their locality. Their answers follow, arranged by states, and with a star in front of such varieties as are considered by the writers as being especially suitable for strictly market or shipping purposes ; "P" means pistillate, and "S" staminate :

PLATE VIII

ROUGH RIDER

PRESIDENT

Arkansas: The following varieties of strawberries are commonly planted in this section: * Aroma (S), * Haverland (P), Michel's Early (S), * William Belt (S), Crescent (P), Klondike (S), * Texas (ɔ), Jessie (S), Lady Thompson (S). These do well here, but occasionally the fruit is killed by the late frosts. It seems that the late varieties are here as liable to this as the early ones. Last season we had a killing frost on the 1st of May which almost entirely destroyed late varieties. Michel's Early always has some berries. The period of blooming is more extended than most varieties. Texas is also an early berry that is given to this, and it bore two crops here two years ago. However, the same season we also had an early and a partial late crop of Michel's Early. Aroma is my favorite late berry; being very firm it is an excellent shipper; and the berries of the last picking are often as nice as the first. Klondike and Texas have quite a name here; many are setting these varieties and doing away with Michel's Early and other smaller-sized berries.—E. H. HALL.

California: Brandywine (S) and Arizona Ever-bearing (S), are mentioned favorably in a list sent in by E. J. WICKSON.

Florida: The best are * Excelsior (S) and * Lady Thompson (S). They bear fruit four to six months.—REASONER BROS.

Iowa: * Gandy (S), Haverland (P), * Senator Dunlap (S), * Warfield (P), Parker Earle (S), Bederwood (S), * Crescent (P).—GEORGE W. STEPHENS.

Kentucky: Warfield (P), Bubach (P), Brandywine (S), * Gandy (S), * Aroma (S).—THOMAS G. FULKERSON.

Maryland: * Bubach (P), Brandywine (S), * Gandy (S), * Warfield (P).—ROY BOBET.

Michigan: * Pride of Michigan (S), * Senator Dunlap (S), * Warfield (P), * Brandywine (S), * Haverland (P), * Sample (P), Climax (S), Enormous (P), Texas (S), Michel's Early (S). The last four varieties mentioned are early kinds and are not so popular as the six later kinds first given, owing to the fact that in northern sections late spring frosts are liable to affect early blossoms. In localities farther south the demand for the early varieties would doubtless rise above the most popular of the medium and late varieties.—W. H. BURKE.

Minnesota: The two leading varieties here are * Warfield (P) and * Senator Dunlap (S). Other kinds do well but are not so prolific.—E. D. FISKE.

Missouri: I have grown thirty-three different varieties of strawberries in the last ten years for market, and this is my latest list: * Michel's Early (S) and * Excelsior (S) for early; * Warfield (P), * Senator Dunlap (S) and * Crescent (P) for medium; * Aroma (S) and * Gandy (S) for late.—A. L. SMITH.

New York (Long Island): We are now in the midst of a big number of tests. So far * Wm. Belt (S) seems to be far and away ahead of everything else. A close second is the * Pride of Michigan (S). * Marshall (S) does exceedingly well with us, and is the most popular with the commercial grower. The high-class trade that seeks Long Island products does not care for Bubach (P) or Gandy (S). Nick Ohmer (S) is grown to a large extent and is a good one. Sharpless yields big and fine-looking hollow mockeries. It is down and out practically all over the Island. Aroma (S) has been but little experimented with, but promises well, and seems to be destined to be a favorite. Bederwood (S), like most early fellows, is hardly worth while here. We find that the so-called mid-season berries come in so close to the extra-early of other territories that I don't think any of the extra-early poor-quality berries will ever get much of a hold with us.—H. B. FULLERTON.

North Carolina: Bubach (P), Climax (S), Excelsior (S), Gandy (S), Lady Thompson (S), Nick Ohmer (S).—W. N. HUTT.

Pennsylvania: Farm Journal's favorite list for this state is as follows: * President (P). * Wm. Belt (S), * Gandy (S), * Sample (P), * Nick Ohmer (S), * Haverland (P).

Texas: * Excelsior (S), * Klondike (S), * Lady Thompson (S), Aroma (S). These varieties ripen in the order given. Generally speaking, this country requires varieties of berries that can stand extreme drought and sudden changes of temperature.—J. E. FITZGERALD.

Wisconsin: * Warfield (P), * Senator Dunlap (S), * Gandy (S), * Sample (P), Haverland (P), Bederwood (S).—J. L. HERBST.

Other varieties: Besides the well-known kinds mentioned

by the foregoing list of correspondents, there are some others that are worthy of mention here. Among these are the following : Clyde (S), Lovett (S), Tennessee Prolific (S), Ridgeway (S), Glen Mary (S), New York (S), Rough Rider (S), Chesapeake (S), Gov. Fort (S).

Order of ripening : A well-known Michigan grower furnishes the following list, embracing nearly all of the standard kinds :

EXTRA EARLY	EARLY
Excelsior (S)	Bederwood (S)
Climax (S)	Clyde (S)
Michel's Early (S)	Lovett (S)
Texas (S)	Tennessee Prolific (S)
	Crescent (P)
	Warfield (P)

MEDIUM	LATE
Lady Thompson (S)	Aroma (S)
Ridgeway (S)	Pride of Michigan (S)
Glen Mary (S)	Brandywine (S)
Wm. Belt (S)	Gandy (S)
Klondike (S)	Marshall (S)
Nick Ohmer (S)	Parker Earle (S)
New York (S)	Rough Rider (S)
Senator Dunlap (S)	Bubach (P)
Haverland (P)	Sample (P)
Enormous (P)	
President (P)	

Fall-bearing varieties : Several kinds of strawberries have shown an ability to bear fruit in the late summer or fall months. Pan-American (S) and Autumn (P) are the best known of these kinds. If a berry grower will plant these varieties, give them the same culture as other kinds, and then keep off all blossoms until about August 1st or even later, he may hope to obtain a satisfactory crop of fall berries. Permit me to say, however, that Pan-American makes very few runners, and therefore the supply of this variety is limited and rather high-priced. Autumn is a better plant maker, but being a new variety, the plants are expensive. Owing to the flood of other fruits in the fall, I am inclined to doubt, from a market standpoint, the desirability of *many* fall-grown strawberries. In a small way, or as a curiosity, they are all right. Some growers propagate Pan-American plants by division of the crowns in early spring, and in this way partially make up for the lack of runners. Americus and Francis are in the class of promising new fall-bearing varieties.

Ever-bearing strawberries : The foregoing varieties produce berries from June until frost, with perhaps a pause now and then. The late crop, however, is apt to be smaller than the early crop, hence the advisability of keeping off all early blossoms if you want a good *fall* crop.

New varieties : The old proverb says, "Try all things and hold fast to that which is good." And the idea is sound. But you should not waste much money or space or time on the "trying"—let the State Experiment Stations do that, and then you read their reports. It is all right for the average strawberry grower to try a *few* novelties each year, but for the main crop he should bear down hard on standard, time-tested kinds. There is great activity among berrymen to originate and introduce new seedlings, and I am glad that it is so, for this is a worthy work and must result in great good. If they will but give us one variety of merit annually, their enterprise will be justified and they will deserve the thanks of their generation, so I wish them abundant success in their endeavor. In this book, however, I have not attempted to list the many newer varieties which have very lately come upon the market. A few of them, doubtless, have merit, and in time they may perhaps replace some of the kinds that are now considered standards—just as the old Wilson aud Sharpless have been replaced by later introductions.

"Freaks": I call a "white strawberry" a freak—don't you? And who wants freaks, anyhow? Also a "white blackberry" is a freak. I object to such reversals of nature's colors, just as I would object to pink bluebirds or black snow. Such things can serve no useful purpose.

PLATE IX.

CUTHBERT

LOUDON

MILLER

CHAPTER VII

THE RASPBERRY

Thorns are Dame Nature's needles, and raspberries are her thimbles.—Dorothy Tucker.

Having devoted a fair portion of this book to the strawberry, I now come to the other small fruits, — fruits of great economic importance, for, with the strawberry, they form an unbroken succession of highly palatable and wholesome food during the entire summer, and are quick sellers in the markets.

Referring to my garden diaries of past years I find that the strawberry season, in my individual case, extends from May 26 to July 3, the raspberry season from June 27 to July 21, the dewberry season from July 4 to July 20, the blackberry season from July 16 to August 22, and that I cut grapes for market sometimes as early as August 20. This shows how one fruit overlaps the season of its successor.

These dates are not extreme, even for my own neighborhood, for somebody with especially favored location is sure to have berries sooner or later than I can produce them. One neighbor, for instance, has strawberries a week after mine are done bearing, on account of his situation on a northward-sloping hillside. The quoted dates are merely suggestive.

The raspberry occupies an important place in the succession of small fruits, and there would be a

serious break without it.   Its culture is easy.   It is a
sure cropper under good treatment, excellent as a
table fruit after strawberries are gone, and sells well
in the markets.

SOIL.—In setting out a raspberry patch it is well
to select a deep, loamy, well-drained soil, and to en-
rich it properly.   Blackcaps require richer, heavier
soil than red varieties.   Too much stable manure
causes the reds to go "too much to canes."

PROPAGATION.—To get a start, buy plants of
a nurseryman ; or propagate from an old patch as
follows :   New plants of the red raspberry may be

obtained by dig-
ging the larger vig-
orous roots and
cutting in pieces
two or three inches
in length, accor-
ding to their size;
the smaller the root
the longer it should

CUTHBERTS ARE POPULAR EVERYWHERE

be cut.   Cut the roots in the fall and store in boxes
of sand placed in a dry, cool cellar until spring.
As soon as the ground can be properly prepared,
scatter the root pieces thinly in furrows and cover
with two inches of light, loamy soil.   Choose a moist,
partially-shaded situation, keep clean and free from
weeds, and by fall you will have a good supply of
strong, healthy plants for early spring setting (for
the North I favor spring setting).   An easier way is
to dig suckers or sprouts that come up along or
between the rows, being sure to secure with each

sprout a short portion of the cross root from which it grew ; dig and set these in permanent rows in the early spring. Much of this digging, however, hurts a patch.

Blackcap raspberries do not sucker from the roots and are propagated differently. Toward autumn when blackcap tips bend down near the ground, new plants can be easily started. Bend down and bury each tip a few inches beneath the ground, holding it in place by pegs, a stone, or the weight of a little heaped-up soil. Most of the tips, if not disturbed, will take root and form nice plants by next spring ; at which time the parent canes can be severed a few inches from the new plants, and the latter can then be dug up and set out wherever desired.

PLANTING.—Red raspberry rows for horse cultivation are usually made about six feet apart, plants spaced about two feet apart in the row (3,630 to the acre). The plants sucker and run together in the

row in a year or two, until there is a continuous hedge-row about a foot wide; plants which come up outside of this, or feeble or surplus plants, should be treated like weeds. For small garden or hoe cultivation the rows might be a little closer together— say five feet.

Blackcaps for horse cultivation may be set in six-foot rows, about two and one-half

PLANTING RED RASPBERRIES—
ROWS ABOUT SIX FEET APART

feet apart between plants. Or they may be set 5 x 5, and cultivated both ways if the rows are straight in each direction (1,742 plants to the acre). As black-caps do not sucker, the hills will ''stay put.''

As to the depth to set raspberry plants, I shall

simply say : Set them only a trifle deeper than they were before digging.

CULTIVATION.—This should begin in early spring and continue, say at ten-day intervals, until about the 1st of August, when all culti-vation should cease so as to allow the canes to stop growing and harden up for the win-ter. Later cultivation would mean later growth, more tender canes, and greater like-lihood of winter-killing. A mulch at fruiting time is sometimes help-

HOE AND CULTIVATE RASPBERRIES
UNTIL AUGUST

ful and practicable in a small patch.

A cover crop is sometimes sown at the last culti-vation, for turning under in early spring. This is often a good idea. Crimson clover, winter vetch, rye, oats, etc , are used for this purpose.

Do not plow the ground, after the raspberry

plants are set, deeper than three inches ; cultivate about two inches deep ; hand-hoe between plants where the cultivator can not go.

Do not let plants produce fruit the first season ; a small crop may be expected the following year ; a full crop the third year.

MULCHING.—A grower in New York state has this to say on this subject : "I wish to call your attention to the advantages of a system of mulch treatment over that of cultivation in the care of raspberries, etc. For many years I have depended on heavy mulch alone, applying coarse, strawy manure, refuse from the lawn and garden, etc., for that purpose, to a depth sufficient to keep down growth of grass and weeds around the bushes, and with the most satisfactory results. The yield and size of fruit are increased, especially in a dry season, and the length of profitable, bearing age is also considerably extended by this method. I have had individual plants continue to produce heavy crops of fruit for ten or more consecutive years when thus treated. Try it and be convinced of its merits."

Remarks : This mulch method may be satisfactory in some cases, but in most cases I believe that cultivation is better than mulching. Constant mulching means a harboring place for insects and fungi. A mulch at *fruiting* time, however, is an excellent thing in combination with early cultivation.

PRUNING.—The first year, none. After that, cut out, close to the ground, all old canes each summer *as soon as they have fruited*. At the same time cut out surplus canes, when the rows or hills get too thick, and diseased or feeble canes. Remove and burn the cuttings, promptly. Don't let the rows get too wide or too thick, but be sure to leave *enough* new shoots for next year's fruiting. In the early spring go through the patch again ; cut out all broken or winter-killed canes or branches, shorten remaining canes to four or five feet, and cut off at least a third of the tips of long side-shoots. (Note : In the North this secondary pruning should not be done in the fall or winter.) Rake up and burn all brush. The best

FIG. 1

pruning tools to use are a long-handled hook (Fig. 1) with which to cut off canes close to the ground, and a pair of ordinary pruning shears (Fig. 2) for work higher up.

There is another pruning detail which is practised by some growers, called summer pruning or pinching.

This consists of pinching off the tip ends of all new canes when they are not more than two feet high— the idea being to make the canes stocky and more self-supporting, with low side-branches. This method has advantages and disadvantages; some growers favor it, while

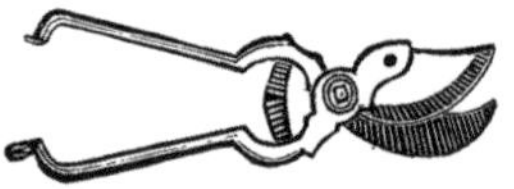

FIG. 2

many others condemn it and say that it often causes too much late, tender growth that winter-kills in the North. It seems to work better with blackcaps and blackberries, than with red raspberries.

If the canes are properly pruned, no supports should be needed ; although in small gardens it is quite common to string stout wires along the rows,

PLATE X.

KANSAS

GREGG

CONRATH

from end posts, either using double wires or tying the canes to a single wire.

INSECTS AND DISEASES.—Cane-borers, gall-beetles, tree crickets and similar insects that infest raspberry canes are difficult to combat with sprays; however, the prompt cutting out and burning of old, dead and infested canes will usually keep these enemies in check. A little worm, the larvæ of a black saw-fly, sometimes feeds upon the leaves; hellebore or arsenate of lead sprays

THESE RASPBERRY CANES WERE PRUNED RIGHT AND NEED NO SUPPORTS

will kill it. Anthracnose (purplish or scabby patches on the canes) is a fungous trouble; spray with the Bordeaux mixture and promptly destroy canes after fruiting. Red rust (powdery, orange-red places on leaves, etc.) is a very common trouble; dig out infested plants—root and branch—whenever seen, and burn; be careful not to scatter the dust on healthy bushes; early sprayings with Bordeaux may help a little.

AFTER-CARE.—Annually the ground should be fertilized with well-rotted stable manure, applied along the rows, supplemented with a generous application of ground bone and wood ashes or ground bone and muriate of potash. The ground bone may be as much as 600 pounds to the acre, and the muriate of potash 200 pounds to the acre, in addition to the stable manure. To fail in the matter of fertilizing raspberries is to bid for small sized fruit. Plow the patch each year as soon as the ground is workable in the spring, and cultivate or mulch it until August.

It is imperative that raspberry patches be moved every four or five years, for best yields. Eradicate the old patch and put the ground into some other crop for several years.

Winter protection : In very cold climates, and especially in the case of some varieties which are more or less tender, it is sometimes advisable to protect the canes during the winter. This is usually done by bending them down along the row and covering them with soil. Remove the covering and straighten the canes in early spring before growth starts.

YIELD.—In reply to the question, ''What do you consider a fair average yield per acre?'' I have figures from fifty-eight growers, says F. W. Card. Computing the average from all these replies, as accurately as possible, I have for the answer 2,493 quarts, or nearly seventy-eight bushels per acre. The majority gave the number of quarts or bushels which they considered an average; others placed their answer in the form of ''from seventy-five to 100 bushels,'' and two gave what they considered high or maximum yields, making it a little more difficult to get the exact average. The lowest estimate given as an average yield was at the rate of 576 quarts; the highest, 9,600 quarts. I judge that neither of these two are extensive commercial growers.

VARIETIES.—There are three types of raspberries —red, black and purple. The yellow forms belong with the reds, and have been derived from them. The reds have a wider range of soil and climate than the blacks. The blackcaps are now largely grown for canning and evaporating. For table use the reds are in most demand. The following suggestions about the most popular varieties in different states, have been sent to me by various correspondents. Stars designate the kinds most preferred by market growers. "R" means red, "B" means blackcap.

Arkansas: The Kansas (B) does well on well-drained soils. The Turner (R) does well on my land, and brings a fancy price, but as the berries are very soft it is difficult to get them shipped in good shape.—E. H. HALL.

California : Cuthbert (R) is the most popular variety, says E. J. WICKSON.

Florida : Raspberries do not do well in Florida, writes a Southern correspondent.

Iowa : *Cumberland (B), *Gregg (B), Kansas (B), *Cuthbert (R), *Loudon (R).—List furnished by GEO. W. STEPHENS.

Kentucky: Cuthbert (R) is the only variety recommended by THOMAS G. FULKERSON.

Maryland: Cumberland (B), Kansas (B), *Gregg (B), *Cuthbert (R).—List sent in by ROY BOBET.

Michigan : *Cuthbert (R), *Gregg (B), Cumberland (B) and Ohio (B).—W. H. BURKE's list.

Minnesota : King (R) and Loudon (R) are E. D. FISKE'S favorites.

Missouri : I grow only blackcap raspberries—Kansas and Gregg.—A. L. SMITH.

New York (Long Island): Of the black raspberries we find two varieties which stand so far ahead of all others under our climatic conditions that we recommend nothing else. The leader is Cumberland, and a close second is Munger. In flavor and size Cumberland easily leads. In yield Munger appears to beat the Cumberland, but it is only in appearance. The great size of the Cumberland easily makes up for the larger number of berries the Munger produces. In red raspberries we find Cuthbert leads all the rest.—F. B. FULLERTON.

North Carolina: Eureka (B), Gregg (B), Kansas (B).—W. N. HUTT'S list.

Pennsylvania: * Gregg (B) and * Cuthbert (R) seem to be the favorites in this state.

Wisconsin : * Gregg (B), * King (R), Cuthbert (R).—List furnished by J. L. HERBST.

Yellow varieties: There is one yellow raspberry well worth raising, mainly because it sets off so beautifully a dish of either the black or the red varieties. This is the Golden Queen, which with us is the best; when thoroughly ripe it is exceedingly good in flavor, but is not quite so sweet as either the black or the red.—H. B. FULLERTON. (Note: This variety is propagated, pruned and grown in the same manner as the red raspberry.— J. B.)

Purple varieties : Shaffer and Columbian are the two best-known kinds.

Propagation and culture the same as for blackcaps. Purple raspberries are not generally popular in market, owing to their unattractive color ; but they are sometimes planted for home use or canning.

Other kinds : Besides the standard varieties already mentioned in this chapter, there are a host of others of different degrees of merit, among which are : Marlboro (R), Miller (R), Brandywine (R), Conrath (B).

I have not space to mention the many new kinds which have not yet been thoroughly tested. Doubtless some of them may some day be counted among the " standards."

Chapter VIII

THE BLACKBERRY

There is no bush fruit which is capable of yielding greater profit.—Prof. L. H. Bailey.

While anybody may grow blackberries, nobody should do so who does not intend to take care of them, for a neglected blackberry patch is as much of a wilderness as a piece of wild thicket land. Besides, disease hostile to good berries lurks in decaying canes and dead leaves. The patch must be pruned, cleaned, cultivated, and kept in good order.

The blackberry has a true place and a high place in the list of small fruits, for if picked only when fully ripe it is a grand table berry, and if grown properly the yield per acre may reach 250 to 300 bushels, which means anywhere from $150 to $300. An average yield of blackberries, however, is said to be 3,158 quarts, or about ninety-eight bushels.

ELDORADO BLACKBERRY
(One-half size)

Blackberries are adaptable to many soils, but do best in a deep, mellow loam, abundantly supplied with humus. They will, however, thrive on soil that is too light, dry and poor for raspberries or strawberries; but they do not like "wet feet," and appreciate a fair amount of fertilizers. Too much stable manure (or nitrogen) is not advisable, as it tends to make too rank a growth of canes. A little nitrogen and considerable potash and bone meal make an ideal food supply for blackberries.

PROPAGATION.—The same as advised for red raspberries (see Chapter VII).

PLANTING.—Blackberries for horse cultivation are usually set about eight feet apart in rows, plants

spaced about two feet apart in the row (2,722 to the acre). For small garden or hoe cultivation the rows might be a little closer together.

The plants, as do red raspberries, sucker and run together in the row in a year or two, until there is a continuous hedge-row about fifteen inches wide. Plants which come up outside of this should be destroyed with the hoe or cultivator.

DIGGING A BLACKBERRY PLANT FOR NEAR-BY SETTING

CULTIVATION, PRUNING, DISEASES, ETC.—Read the directions for red raspberry culture given in the preceding chapter. Blackberry pests, culture, etc., are similar.

Some blackberry growers stretch a wire along the row, about three feet from the ground, to which the canes are tied. Two wires may be used, one above the other (see Fig. 1), the long canes being tied and

treated like grape-vines. Or, the two wires may be
placed side by side, say three feet above ground, and
the canes required to stand
between the wires. Large
areas of canes are seldom
supported in any way; and,
generally speaking, I do not

FIG. 1

think it is necessary in any case if the pruning is
properly done.

The blackberry patch should last for a score of
years, and more trouble and expense are therefore
warranted than in the case of a transient crop like
strawberries. The end in view in blackberry culture
is to keep the ground under good tillage ; to keep the
rows clear of dead-wood and trash ; and to facilitate
the gathering of the crop. The work of pinching back
or nipping (see Chapter VII) the growing canes, if
done at all, is more easily performed if the rows of
canes are kept narrow and compact.

It is essential to harden the young wood by ceas-
ing culture in August. The cultivator should run
frequently and regularly during the spring and early
summer.

In my latitude the last cultivation will occur about
August 20th to 25th, after which no more encourage-
ment should be given the canes in the direction of
growth. The entire autumn is thus given for matur-
ing the wood made by the young canes, and my
canes seldom suffer from winter killing.

Hardy varieties are preferable to those which are
tender ; but where the necessity for winter protection
exists it is easy to remove the earth from one side of

a bush or bunch of canes, force the canes over into a reclining position, and cover them with soil.   Where

SPRING PRUNING SHOULD BE DONE
BEFORE LEAVES ARE FULLY OUT.
THIS MAN IS A LITTLE LATE.

this is done the canes must be liberated in early spring, as soon as danger of severe freezing is over.

As to growing supplementary or what are called hoed crops in young plantations of raspberries and blackberries, the question is one for the individual operator to decide.   It will perhaps do no harm to put in a row of something in the middle of the space between the rows of berries, but this cropping should be done only the *first* year, after which time the canes and their roots should have all the space in the rows and aisles.

ESTIMATE OF EXPENSE. —In a cold climate where canes must be protected in winter, the following estimate for one acre of blackberries was made after many years' experience on the ''Thayer Fruit Farms'' and indicates methods adopted in Wisconsin :

Plowing land . . . . $ 1.50
Harrowing, 4 times 2.00
Marking and laying out 1.00
Plants 30.00
Setting plants 5.00
Cultivating, 15 times 7.50
Hoeing, 3 times 3.75
Manure, 20 loads for mulching 15.00
Covering plants, for winter 2.50
    Total expense, first year $68.25

Removing covering 2.50
Cultivating, 15 times 7.50
Hoeing, 3 times 3.75
Plants, and resetting missing hills 8.75
Nipping and pruning 2.50
Mulching and manure 25.00
Posts for support, 62 . . . . 4.00
Stakes for support of vines, 300 . . . . . 6.00
Wires for support, 300 pounds No. 12 . . 9.00
Labor on support 3.75
Laying and covering for winter 5.00
Use of tools 4.00
    Total expense for two years $150.00

VARIETIES.—The following kinds are favorably mentioned by correspondents in different states. A star means "specially adapted for market purposes":

California: Crandall, Lawton, Kittatinny, Himalayan —E. J. WICKSON'S list.

Indiana: In 1883 we planted one-third of an acre in three varieties,—Early Harvest, Ancient Briton and Collins. The ground was well cultivated for several years. After plants were well established, all that was done was to remove all old wood and rubbish and occasionally mulch the ground. From 1887 till now this patch has given greater net returns than any fruit ground we have. In 1907 this twenty-four-year-old patch out-yielded all preceding years to so great an extent as to border on the wonderful. We picked the first berries from the Early Harvest rows on July 9th and continued daily pickings until September 1st, a period of fifty-four days. On September 1st we gathered a few quarts of Early Harvest, making this variety king of bearers over all other varieties on our place. The yield of this one-third of an acre was 1,327 quarts that season.

Total receipts	$128.20
Commission	12.80
	$115.40
Paid pickers	17.75
Net receipts	$97.65

The season was unusually moist, which kept up the vigor of the plants to the last. At this ratio an acre would net about $290. The next season (1908) was unusually dry, and so the returns were very much reduced, being only $45; crop less, but prices better.—J. H. HAYNES.

Iowa : GEORGE W. STEPHENS recommends just two kinds—* Snyder and * Ancient Briton.

Kentucky : THOS. G. FULKERSON mentions only one kind—Early Harvest.

Maryland : * Illinois, Early Harvest and Snyder is the list favored by ROY BOBET.

Michigan: Snyder, Erie and Rathbun, says W. H. BURKE.

Minnesota: Ancient Briton and Snyder are the favorite team of E. D. FISKE.

Missouri: I can sell in my home market more Early Harvest than all the other varieties combined. Kittatinny is a good berry if you are willing to fight the rust.—A. L. SMITH.

New York (Long Island): Eldorado, Rathbun and the old Lawton are "nip and tuck."—H. B. FULLERTON.

Ohio: Eldorado and Erie.

Texas : * McDonald, * Dallas, Early Harvest, Lawton, Kittatinny, Snyder. These are given in the order of ripening. —J. E. FITZGERALD.

Wisconsin : * Ancient Briton and * Eldorado.—J. L. H.

PLATE XI.

CUMBERLAND

BLACK NAPLES CURRANT

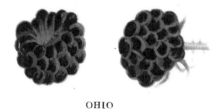

OHIO

CHAPTER IV

THE DEWBERRY

Too rambling to suit me, but a splendid berry.—Tim.

The dewberry is distinguished from the black-berry chiefly by its trailing habit, says J. Troop, of Indiana, and its method of propagating by tips instead of suckers. It is found growing wild in many parts of the country in woods and fields that have been more or less neglected. The fruit, however, of these wild forms is generally too small and too poor in quality to make it at all desirable for cultivation.

The dewberry's chief value lies in its season of ripening, which is just between raspberries and black-berries.

AFTER PLOWING AND HARROWING, USE THE HOE

CULTIVATION.—Dew-berries are cultivated about the same as blackberries, except in the matter of training the vines. On good moist soil the canes will often make a growth of ten feet or more in a single season. Even on poor, light soil,

dewberries grow remarkably well. The canes may
be trained to stakes or wire trellises ; or they may be
heavily mulched with clean straw, in order to protect
the fruit, and allowed to run at will. The latter
method will require much more room, and is not so
satisfactory as tying to stakes, when it comes to pick-

ing the fruit. The old fruiting
canes will need to be cut out
each season, and the young
ones cut back quite severely
in order to get the best
results.

Allow only four or five
canes to grow in each hill.
For the home garden, a
dozen ...nts should furnish
all t.. fruit an average
family will require.

A Maryland grower cul-
tivates forty acres of this

TYING UP DEWBERRIES—ONE
VINE TO A STAKE

fruit, and one year sent 45,000 quarts to market.
He began picking June 20th and ceased July 12th,
covering a period of more than three weeks, and
reaching the market well in advance of the main crop
of blackberries ; and he profited accordingly. He
feeds his land well, and puts it in good mechanical
condition. He sets the young plants (rooted tips,
obtained in the same way as blackcap raspberry
plants) three feet apart in rows six feet apart. This
requires 2,420 plants to the acre. The vines are not
tied up the first year, but are simply kept well cul-

tivated and allowed to run. Early the following spring stout stakes are driven along the rows in such a manner that two plants may be tied to each stake; that is, there are just half as many stakes as plants, the stake standing midway between two plants. The stakes are about three feet high, after being driven into their places, and the vines are stretched straight and tied only to the top of the stake. Each pair of vines occupies an A-shaped position (see photo-engraving). This is the bearing wood of the year. The new growth of the season is allowed to

A MARYLAND METHOD OF TRAINING DEWBERRIES

scramble upon the ground in its own natural way. When the crop is off, soon after the middle of July, the old wood is removed, and the strength of the vines thrown into the new growth. The patch is cultivated and cleared of weeds, and made ready for a repetition of operations the following season.

I have measured dewberries that were nearly one and one-half inches long. The vines are very thorny, and he who works among them should protect his hands with leather gloves.

YIELD.—A dewberry plantation will last a good many years, and does not reach its best production until three or four years old. The yield varies greatly from year to year and in different localities. From forty to sixty bushels per acre is perhaps a fair average. Sometimes the blossoms fail to set fruit satisfactorily, which failure is probably caused by lack of proper pollination ; this may be overcome by planting several varieties together in alternate rows.

WINTER PROTECTION.—In the North, where the winters are severe, the vines should be laid down in the late fall and covered with earth until early spring.

DEWBERRY PICKERS SHOULD PROTECT THEIR HANDS WITH FINGERLESS LEATHER GLOVES

INSECTS AND DISEASES.— See Raspberry and Blackberry.

VARIETIES. — Lucretia is the kind which seems to be best liked nearly everywhere. Correspondents in a number of states speak favorably of it. Austin is another good variety, which originated in Texas and is largely grown there ; it does well in Maryland and elsewhere.

PLATE XII.

FAY

FAY

RED DUTCH

CHAPTER X

CURRANTS AND GOOSEBERRIES

Give them just twice as much manure as you think they need.

Currants and gooseberries thrive under a wide range of conditions, but do best when planted in deep, moist, cool soil.   Clay soil, with good drainage, well enriched, suits them almost perfectly.   It is a good plan to mulch around the bushes with straw or with green clover cut in full blossom, through the heat of summer.

Some growers shade their bushes by alternate rows of grape-vines and some by means of fruit trees, for these bushes seem to like partial shade.   On large, commercial plantations in the North, however, no attempt is made to supply shade.   Shade is more necessary south of the Ohio and Potomac rivers than in more northern latitudes.

Do not stint the manure.   These fruits require extra-heavy manuring in order to grow berries that will command the best price.   Not only should the soil be in excellent tilth at the time of planting, but it should be top-dressed yearly every autumn or early winter with pig or cow manure ; also, bone meal and potash should be applied occasionally.   There are no fruits that will respond more quickly to good treatment than currants and gooseberries.

PLANTS AND PLANTING. — The easiest way to get a start is to buy plants of a nurseryman and set them either in the early spring or in the fall, preferably the former, unless you are willing to mulch them well through the winter.   Or you can grow your own plants for setting, as follows:

Soon after the leaves fall in autumn, currant and gooseberry (or grape) cuttings may best be made.

Use this season's new wood - growth ; cut into lengths about eight inches long, tie into bundles, and bury in not too wet sand in the cellar, for spring planting.   Or if you will mulch them well through the winter, they may be set at once in the fall.   Set them slightly leaning, top end up, about five inches apart, deeply, in rows, each cutting having at least one bud above and one below

SHALLOW CULTIVATION IS BEST
FOR THESE BERRIES

ground ; cultivate them as you would any other crop, and they should be ready to transplant to their permanent place the following fall or spring.   The soil should be well firmed around the cuttings, and a little shade and an occasional watering may not be amiss while they are striking root.

Most growers make the permanent rows 5 x 5 feet and cultivate both ways ; but in a small garden the bushes might be spaced about four feet apart in five-foot rows.   A well-cared-for plantation should last a long time.

CULTIVATION. —The roots run close to the surface, so beware of cultivating deeply ; but regular, light stirrings of the soil until August are beneficial. Some growers cultivate the ground several times in the spring and early summer, and then mulch the surface during the balance of the year ; this is a good hint for the small gardener.

PRUNING AND PESTS. —These fruits are usually grown and trained in bush form, and shoots from the base are utilized to renew the top gradually. Pruning consists of cutting out weak or superfluous new shoots, and old ones which have outlived their usefulness or which have become diseased or infested with borers. But never cut out too many old stalks at once, for the bulk of the fruit is produced on such wood. Stalks may be left until about three years old — which is often the limit of their usefulness. All shoots, old or new, may be cut back, as desired, to make the bushes more compact and symmetrical, and the fruit larger. In other words, thin out, cut

PULL WEEDS BY HAND THAT CAN'T BE REACHED WITH THE HOE

back, and gradually replace the older stalks with younger ones. Burn all clippings promptly, for thus

the borer is kept down. The principal other enemy is the currant-worm, which attacks the leaves of both currant and gooseberry in early spring.

Remedy : Spray the bushes with the hellebore mixture, but do it promptly when the first worm is seen ; watch the bushes closely shortly after the leaves unfold in the spring. Fungous blight troubles can be controlled

CURRANT-WORMS DIDN'T GET THIS PATCH —EARLY HELLEBORE SPRAYS KILLED THEM

with Bordeaux ; it is advisable to use this mixture on all plantations of currants where the foliage drops early. Sulphur for mildew is often used.

In some localities the currant-fly is troublesome. It deposits eggs in the berries and soon the currants are wormy. No good remedies are known. Preventive measures, however, are often helpful. These are : The prompt gathering and burning of infested fruit ; allowing chickens to run among the bushes.

If lice attack the leaves, use one of the lice rem-

edies given in Chapter III.   For the San Jose scale
use the lime-sulphur mixture.

YIELD AND PROFITS.—The average yield of currants has been put down at 2,000 quarts per acre,
with yields reported as high as 7,500 quarts per acre.
Net profits will depend on market prices and expenses,
and both of these items are variable.   A recent New
Jersey bulletin intimates that about $150 per acre net
profit may be expected.   Of course it all depends
upon circumstances, but I am sure that an energetic
man near a good market can do well with currants,
provided he does not undertake too large a patch.
Many of our horticultural operations would be more
successful with acreage divided by two.

Gooseberries are sometimes a paying market crop,
and sometimes they are not.   Some years the markets are fairly glutted with this fruit and prices drop
woefully.   ''Go slow'' on this crop until you feel
your way and find out market conditions in your
locality ; some markets take gooseberries better than
others do.

VARIETIES. — Vic-
toria, Cherry, Fay's
Prolific, Red Cross, etc.,
are large-fruited popu-
lar kinds of market
currants nearly every-
where.   Red Dutch is
the small, old-fashioned
kind ; it is still much
grown.   White Dutch
and White grape are

A QUART BOX OF HOUGHTONS
READY FOR MARKET

good white varieties.   Black Naples is esteemed for jellies, etc.

Downing, Houghton, Columbus, Chautauqua, Pearl, etc., are well-known gooseberries suited to the American climate.   Industry is a fine, large, European variety, very sweet ; but more subject to mildew than the other kinds.   E. S. Holmes, New Jersey, says that the Houghton is the best gooseberry for market ; and reports from other states – as far west as California—seem to confirm this statement.

A Florida correspondent writes that neither currants nor gooseberries are grown there, as a warm climate is unsuited to these fruits.

Chapter XI

THE GRAPE

Nothing great is produced suddenly, not even the grape.
—Epictetus.

To start a vineyard in the North, buy one-year-old vines of a nurseryman and set them in early spring ; or propagate them from cuttings as advised for currants. Vineyard rows should be about eight feet apart, plants spaced about eight feet in the rows. Any good, well-drained soil will do, but—commercially speaking—grapes do their best only in certain localities where temperature, climate, water protection, etc., are entirely favorable. The Chautauqua "grape belt" in New York state is such a place. There are others. However, a few grapes for home use can be grown almost anywhere. On account of greater security against frost dangers, an elevated location is preferable to a low-lying place; so is a location which has a large body of water between it and the direction from where cold winds usually come. In very cold sections grapes do especially well on the sunny side of a wall, fence or building.

BE SURE TO REMOVE LABEL WIRES THAT MAY CHOKE VINES

PRUNING.—After planting, cut back the top to about three buds and let the vine grow as

it pleases the first season.   Then, the next February,
cut back the best cane to about four ''eyes'' or buds,

and cut off any
other canes en-
tirely; when these
buds commence to
grow, rub off all
but the two strong-
est shoots, and, as
they grow, tie
them to the wire
of a trellis or to a

STRINGING WIRE, AND STRETCHING IT WITH
A BLOCK AND TACKLE

stake.   The third year's pruning will depend somewhat upon what system of training you decide upon;
the subject is too large and intricate for treatment
here.   The Kniffen system is one often used; it
consists in training the vine so that it has four
horizontal side or main branches, two on each side,
one above another, tied to
two wires; the first wire may
be about three feet high, the
second about five feet.

There are several other
methods of training grapes
—some requiring only one
wire, some three wires, and
some merely a stake or post
for each vine.   The main
thing is to tie up and support the vines in any convenient way — on wires,
posts, arbors, trellises, build-

STAPLING A STRETCHED WIRE
IN PLACE

ings, walls, or fences — and then cut out the new
wood each year down to two buds on each shoot.
Every grape-grower should send to the Secretary,
U. S. Department of Agriculture, Washington, D. C.,
ask for Farmers' Bulletins Nos. 156 and 284, and learn
all the facts about the various pruning systems and
methods of culture.   These, together with the information and helpful photo-engravings contained in this
chapter, should give you all the facts necessary to
successful grape-growing.   If you live in the South,
ask for Farmers' Bulletin No. 118, entitled "Grape
Culture in the South."

The secrets of pruning are as follows : Remember
that the fruit of the grape is produced on spring shoots
which come from buds on last season's wood-growth.
To leave too many of these buds means too much
fruit of inferior quality and small size ; to leave too
few buds means a scanty crop of high quality and
large size.   A happy medium is to cut back all new
wood of last season's growth to about two buds ; even
then there may be too many buds left on a vigorous,
mature vine,—in which case you can cut out, entirely,
some of the two-bud spurs.   Pruning should be done
(except perhaps at
planting time) only
when the vines are
thoroughly dormant, or
they will bleed at the
cuts.   In severe cli-
mates, February is an
excellent time to prune;
in milder climates the

TYING UP YOUNG VINES TO POSTS

work is often done in November. I prefer February, myself, although the winter climate in southeastern Pennsylvania is not usually very severe.

CULTURAL HINTS.—Extra-fine grapes are often obtained by cutting off inferior bunches after the fruit has set and removing those that crowd others. Cultivation, too, helps to make fine fruit. So do annual applications of bone meal or phosphate, stable manure, and some form of potash. An oversupply of stable manure or nitrogen, however, tends to make grapes run too much to vines.

Another aid to high quality is sacking or bagging the bunches. The best time to do this is when the berries are quite small — not larger than small shot. Ordinary two-pound paper sacks can be used. The mouth of each sack should be snugly wrapped around the stem of the bunch, and se-

THIS BUNCH IS PERFECT BECAUSE IT WAS SACKED

curely held in place by pinning or tying. The tie should be tight, but, of course, not too tight. The sacks protect the grapes from fungous and insect enemies, including wasps, birds, etc., and the bunches thus protected are finer and more perfect. For grape-rot and mildew, begin early and spray the vines with Bordeaux mixture at intervals of about two weeks. The addition of arsenate of lead to the earlier sprayings will kill any insects which eat the leaves. Aphis or lice should be sprayed with one of the lice remedies given in Chapter III; several kinds of sprayers are there illustrated.

Rose bugs are often a serious pest in vineyards. The simplest remedy is to knock the bugs into pans of kerosene, daily. Or try this as a spray : Ten pounds of arsenate of lead mixed with fifty gallons of water.

Bees are sometimes found on grapes, but their presence does not mean that they are responsible for the punctured fruit. Wasps or birds are the real culprits, and the bees simply feed on the grapes which have already been damaged.

Sometimes some kinds of grapes do not fruit well when planted by themselves,—they seem to need pollinating by the blossoms of other varieties. Concord does well by itself, and is a good pollenizer for other kinds.

COMMERCIAL GRAPE-GROWING.—Mr. S. S. Crissey, a New York grape-grower, contributes the following personal experiences in the famous Chautauqua-Erie district. The influence of a near-by body of water to modify climate, eliminating late spring frosts, and holding back fall frosts till the last week in October, has there been too conclusive to admit of question. Mr. Crissey's contribution contains many hints that should be of help to every grape-grower everywhere —amateur or professional. He says: "For some time I have been growing grapes here, and perhaps my expense and receipt account for one recent year, from a four-and-one-half-acre vineyard, may be of interest:

RECEIPTS

2,323 8-lb. baskets (per Grape sellers
    Union) at 13 3-10 cents, net $308.96
892 8-lb. baskets (private orders) at
    16 8-10 cents, net 149.86
500 4 lb. baskets (private orders) at
    9 1-2 cents 47.50
2,850 lbs. waste at $15 per ton 21.38

                        $527.70

EXPENDITURES

Pruning, tying, spraying and cultivat-
    ing 4 1-2 acres at $12 per acre . . . $54.00
3,215 8-lb. baskets at 2 cents 64.30
500 4-lb. baskets at 1 4-10 cents 7.00
Wages : Picking, packing and cartage 71.30
Interest on $450 (assessed valuation)
    at 6 per cent. 27.00
Taxes, 3 per cent. 13.50

                     237.10
Net Profit $290.60

"My varieties are : Three and one-half acres of
Concord ; one-half acre Worden ; one-half acre Pock-
lington, Wyoming Red, Brighton, Martha, Niagara
and Hartford, mixed. A September hail-storm destroyed about two tons of these grapes. The yield

given is not at all
uncommon; sev-
eral of my neigh-
bors had larger
yields. Now I will
give some practi-
cal hints about
culture, etc.

CULTIVATING A YOUNG VINEYARD

"Chautauqua

has three well-defined types of soil.   Nearest the lake a stiff clayey loam; farther back, a gravelly loam; and on the foot-hills, up to the limit of cultivation, a thin shale overlying the original rock.   Grapes on the clay and the shale have a tougher skin and a better shipping quality.   Those on gravel have finer clusters and larger berries.   Varieties : Nine-tenths of Chautauqua grapes are Concord ; and the other tenth is largely made up of two of its seedlings— Moore's Early and Worden."

Cultivation :   "The first two years," continues Mr. Crissey, "the new growth is left on the ground.

At the beginning of the third year a trellis is made of strong end-posts, and lighter posts between these, twenty-four feet apart.   The best trellis has three No. 9 wires.   Only two canes are put up during the first bearing year. Later, as the vines gain strength, four or five canes are put up and tied in place.   The cultivation is, first,

SOME HAND-HOEING IS REQUIRED
CLOSE AROUND EACH VINE

shallow plowing in May.   This is followed by har-rowing and horse-hoeing, and this by hand-hoeing. (A grape horse-hoe is illustrated in Chapter II.) Vineyards must be kept clean of all grass and weeds up to July 1st.   For a cover crop, to be sown about

July 1st, crimson clover is best. This will be turned under May 15th, following. For the first two years of the life of a vineyard, stable manure is good. For old bearing ineyards, commercial fertilizers with a large per cent. of potash and phosphoric acid are the most used. Pruning is done during late winter. Tying is done in April. Active growth of the buds begins about May 10th. The grapes are in blossom June 20th. August 1st the berries are of full size. Car-load shipments of Concords begin about September 10th, and the season

BROADCASTING COMMERCIAL
FERTILIZER IN YOUNG VINEYARD

for outdoor harvest closes the last week in October. Horse-power sprayers are largely used. Shoulder-strap sprayers are also favored. Bordeaux mixture, arsenate of lead, kerosene emulsion, etc., are in common use. Perhaps the most useful all-round spray is the combined Bordeaux-arsenical mixture.''

(Note : In the chapter on Marketing, Mr. Crissey tells the general methods of selling grapes in the Chautauqua-Erie district.—J. B.)

VARIETIES.—There are hundreds of kinds of grapes, but only a few of them are in general cultivation. Correspondents in various states have sent me the following ideas on this subject. A star means ''specially suited for market purposes '':

California: Not less than fifty kinds are grown here for different purposes. None of them are of eastern origin,—they all came direct from Europe. White Muscat, Muscatel, Malaga, Sultana, etc., are good raisin grapes. Tokay and Black Hamburg are excellent for table purposes.—E. J. WICKSON.

Florida: In northern Florida a few bunch grapes are grown, but in a very limited way. In middle and southern Florida the Muscadine sorts are the only kinds really adapted to the climate; these bear very heavily and are the most delicious of all grapes. The best kinds are Meisch, James and Scuppernong. All of the finest flavor.—REASONER BROTHERS.

Iowa: * Concord, Worden, Moore's Early.—GEO. W. STEPHENS'S preference.

Kentucky: * Concord, Moore's Early, Niagara —THOS. G. FULKERSON'S favorite list.

Maryland: * Moore's Early, * Concord, Niagara, Brighton.

Michigan: * Concord, * Niagara, Worden, Brighton, Delaware.—W. H. BURKE.

Minnesota: Moore's Early and Concord are most popular, but not many grapes are grown here.— E. D. FISKE.

Missouri: The old Concord is all that I grow.—A. L. SMITH.

Pennsylvania: * Concord, Campbell's Early, Agawam, Worden, Niagara, Brighton.—S. C. M.

Texas: * Niagara, * Concord, Goethe, etc., are grown, but grapes are not a popular fruit in this country. They don't sell very well.—J. E. FITZGERALD.

Wisconsin: * Moore's Early, * Delaware, * Niagara.— List sent in by J. L. HERBST.

Long-keeping grapes: The following are good winter keepers: Agawam, Brighton, Canada, Croton, Catawba, Duchess, Iona, Jefferson, Lindley, Merrimac, Rebecca, Salem, Vergennes, Wilder. Choice, perfect clusters of grapes may be kept some time by placing them in layers, packed in dry, clean sand, dry sawdust, cork dust, or something similar. Store in a cool, dry, frost-proof place away from all air currents.

Girdling or ringing grape-vines : This consists in removing a ring of bark from the bearing shoot about an inch wide or wide enough so that the bark will not heal over the wood that has been laid bare. The same result is sometimes accomplished by compressing the branch with wire. The explanation of this effect on the fruit is given in a bulletin of the New York State Station as follows : The food materials taken in by the roots pass up through the outer layers of wood to the green parts of the plant. Here new material for growth is formed and the portion that is not needed by the leaves and other green parts passes downward, for the most part through the inner bark, to be distributed wherever it is needed. The wood is not disturbed in the process of ringing, therefore the upward movement of the solutions is not interfered with ; but since the downward passage takes place through the inner bark the flow is arrested when it arrives at the point where the bark has been removed. Consequently the parts of the plant that are above the point where the ring of bark has been removed receive more than a normal supply of food, which tends to produce increase in size and earlier ripening of the fruit.

Tests show that the ripening of the fruit is thus hastened sometimes as much as ten days or two weeks, and the size increased without loss of palatability if picked in good season. In wet seasons the berries tend to crack open and to be too soft for marketing. In some cases the fruit on the ungirdled branches seems to be reduced in quality ; and girdling, if freely and continuously practised, apparently saps the general vitality of the vine. To avoid injury it is advised to treat only those canes which are to be cut away at pruning time and to leave one-half of the canes untreated. As a further precaution it is suggested that girdling be practised only every alternate year

PLATE XIII.

CHERRY

WHITE GRAPE

CHAPTER XII

MISCELLANEOUS SMALL FRUITS

Experiments are like ice-cream—interesting, but poor as a steady diet.—Harriet.

With novelties the practical farmer or gardener need have but little to do. Most of them are worthless for business purposes. Still, I think it is worth while to keep an eye upon them and perhaps experiment with them a little. The following list of standard berries and novelties is necessarily brief and more or less incomplete ; I have not space even to mention all of the oddities.

BLUEBERRY OR HUCKLEBERRY. — These well-known berries belong to the genus *Vaccinium*, and flourish in a wild state in many parts of the country. The high-bush variety does best on wild moist land and seems usually to object to garden cultivation. The low-bush kind is mostly found on poor, dry land. By burning over these places early in the spring, the bushes are renewed and underbrush kept out. About one-third of a patch is burned over each year. Usually this is all the care given.

BUFFALO BERRY.—This is *Shepherdia argentea* of the botanists. It is a pretty, ornamental shrub, prolific, and highly prized for its fruit in the drier portions of the Northwest. The fruit is small, acid, scarlet in color, with small seeds.

CRANBERRY. — A native of swampy or boggy
marsh land. The berries are borne on low, trailing
vines late in the fall. There are large commercial
plantations at Cape Cod, and in certain localities
in New Jersey, Wisconsin, etc. If any of my readers
are interested in this subject they should write to the
Secretary, U. S. Department of Agriculture, Washington, D. C., and ask for copies of Farmers' Bulletins

SETTING PLANTS IN MANURED FURROWS

Nos. 176, 178 and 221. These bulletins contain full
information about culture, varieties, and insect and
fungous enemies.

GARDEN HUCKLEBERRY. — A novelty offered by
some nurserymen. A Michigan woman, writing to
the Colorado Experiment Station, says of this berry:
"I have grown the garden huckleberry for the last
four years. We are very fond of it for sauce and
pies. It is very nice and nearly like the real huckle-

berry when cooked, but it is not good to eat out
of the hand for it is like eating a green tomato.
Even when the berry is ripe, if you will notice, the
large berry is shaped like a tomato.   The vine looks
a great deal like the nightshades, but if you will
compare the two plants you can see quite a difference.

" The nightshade berry hangs down and the
huckleberry turns up to the sun.   The huckleberry is
larger and different in shape from the nightshade
berry ; and the huckleberry is not poison as is the
berry of the nightshade.   Last year we had fifteen
plants and from the fifteen plants I gathered sixty-
eight quarts of ripe fruit, and much of the fruit was
not ripe last year owing to early frost.   The garden
huckleberries will do well wherever tomatoes will
grow ; they are just as easy to grow as tomatoes.
I received ten cents a quart for one bushel, and
could have sold fifty.   I expect to raise twenty-five or
thirty bushels this year, if frost does not prevent me.
We live in the northern part of the state and some
years we get frost here long before other parts
of the state.''

Personally, I can not recommend the garden
huckleberry, for 1 have not grown it and do not
expect to do so as long as there are better berries
that can be grown.

GOUMI.—Widely advertised under the name
Elæagnus longipes (pronounced lon-gi-pees).   The
word Elæagnus is the botanic genus, and the word
longipes means long footed or long stemmed, referring to the fruit. Goumi is the Japanese name for it.

I am inclined to look with some favor on this

fruit, but can not advise anyone to plant it, except in an experimental way or for ornamental purposes.

Prof. Bailey, of Cornell, says : "It is a graceful and handsome bush five or six feet high, bearing a profusion of silver-white leaves and most abundant crops of cinnabar-red and gold-flecked berries. Whether considered for ornament or for fruit, it is

A FRUIT-LOVER'S BACK YARD

one of the best of the many excellent shrubs which have come to us from Japan." It is perfectly hardy.

JUNEBERRY.—The Juneberries are descendants of our native shadbush, *Amalanchier*. They are catalogued by some nurserymen, but belong in the group of novelties, and have not demonstrated their right to a place among our standard small fruits.

LOGANBERRY. —This berry has, I think, come to stay. It appears to be worthy of the attention of market men in some localities, although the testimony on this point is yet meagre.

This fruit was originated in California by Judge J. H. Logan. Its first bearing was in May, 1883. Its ancestors were Aughinbaugh, a pistillate dewberry, fertilized by "an old variety of red raspberry . . . resembling the Red Antwerp."

The Loganberry is commonly described as being a cross between a blackberry and a raspberry. Its habit of growth is somewhat like the dewberry, and its method of multiplication resembles the blackcap raspberry, as the canes root at the tips. In severe climates it requires winter protection.

ONE WAY OF MAKING FURROWS FOR PLANTING

The fruit is of a highly desirable size and character, partaking of the nature of both parents. It has been called a red blackberry, but has a distinct raspberry flavor.

MAYBERRY.—Novelty. Said to be a promising candidate for public favor ; a member of the raspberry group.

MULBERRY.—Offered in the catalogues, but nowhere very largely grown for market purposes. The Downing mulberry has real merit, but is not quite hardy in very severe climates.

STRAWBERRY-RASPBERRY.—The Rhode Island Station, after a trial, calls this "a veritable weed, entirely destitute of desirable qualities for market purposes." Still, it is a handsome ornament, if nothing more. W. Paddock, Colorado, makes some plain statements about it :   "It is not new ; it is not valuable for its fruit ; and instead of being a cross between the strawberry and the raspberry, it is a distinct species.   This species has been grown in America, in a limited way, for a great many years, and was reintroduced from Japan, where it is native, about twenty years ago.   It was quite widely disseminated a few years later, but it has never developed any commercial importance.   The plants are attractive in an ornamental way, as they make a dense mass of foliage, and flowers are produced through a long period.   The berries are large, red in color, and quite apt to crumble, and they are dry, seedy and insipid.   The plants are unusually unproductive, their fruit-bearing habit resembling the wild thimbleberry of the foothills, and as a commercial sort they have been no more profitable."

WINEBERRY.—The Japanese wineberry has been widely distributed over the country, and has some friends, but does not appear to find public favor for market purposes.

·PLATE XIV.

· VICTORIA

CHAPTER XIII

PICKING AND PACKING

Don't let carelessness now balk a whole year's care.—Tim.

We come now to an important branch of our subject, for gathering and preparing the harvest is half the battle.

I formerly used tickets or cards, containing numbers, and a punch, to keep accounts with pickers, but the past season I tried the system recommended by John M. Stahl and liked it so well that I would not think of returning to the old way. It works like a charm, the pickers are satisfied, and it is no trouble.

A bulletin board is erected just outside of the door of the receiving and packing room. For each day a paper is prepared, to be tacked on the bulletin board. Heavy book-paper of the required size can be obtained at almost any job printing establishment. Or you can use ordinary wrapping-paper. This paper you can rule with horizontal lines half an inch apart. Along the left margin have a vertically-ruled space for the numbers; next have a space for the names of the pickers; and then a dozen or more spaces in which to put down the number of quarts brought in by each picker. (The accompanying illustration, Fig. 1,

FIG. 1

is merely suggestive ; there should be more spaces or columns for "quarts brought in." The right-hand space is for figuring totals at the end of the day.) Every picker has a number. This is important ; let the pickers be referred to by their numbers, not by their names.

When each picker brings in a load, the number of quarts is marked in a space opposite the number of the picker. If an indelible pencil is used the pickers can not accuse you of altering the record ; and if you put in the number of quarts in the presence of the picker, there will be no oversights or mistakes. The entire record is open to any picker at any time during the day when he or she comes to deliver berries. You can see at a glance how each picker is working ; or, if you desire to know at any time how many quarts have been brought in, you can foot it up in a minute.

Each evening the record sheet is taken down, folded, and the date, number of quarts picked, and whatever other memoranda may be desired, are endorsed upon it. It is then filed away. These sheets furnish a complete account of the season's picking. They also furnish valuable information for future use.

I have found it advantageous to supply each picker with a berry tray or "carrier," on which the boxes, when filled, are borne to the packing shed. My trays were made according to the following directions, and seem well adapted to the service required of them : For the ends use inch strips three inches wide; for the bottom, four strips of laths; and for each side, one strip. (Some growers add four legs to the tray, so as to raise it off the ground when in use.) A handle is made from half of a barrel-hoop, spanning the tray lengthwise, and tacked to the end pieces on the outside. This tray is designed to be made large enough to hold six one-quart boxes.   Placing the handle lengthwise, instead of crosswise as shown in Fig. 2, leaves the boxes easier to get at, and prevents the tray from tipping.   I use these trays only to put the boxes in after

FIG. 2

the pickers fill them, and not to pick in; although I know that many growers have the pickers carry them along while picking, — but this often jostles and injures the fruit, exposing it to the evil effects of the hot sun, and weights the picker.   Near the middle of the day, especially if the sun be hot, it is best, after filling a box, to set it among the foliage, hidden from the rays of the sun, until a trayload is picked, and then carry all, in the tray, to the packing shed.

J. H. Hale, Connecticut, sums up his strawberry picking and packing methods as follows: "If wanted for local market, start picking at daylight, and have pickers enough so the fruit can be gathered and into the market before eight o'clock.   For distant market, try to pick in the evening, or in the morning *after the dew is off* and yet before it is too warm. If picking must be done all through the heat of the day, plan some way to cool the berries.   Pickers of mature years are best ; and, as a rule, girls are better than boys.   Have a superintendent for every ten or twelve pickers, to assign the rows, inspect the picking, etc. Each picker should be numbered and have a picking stand or carrier with like number to hold four, six or eight quarts. Sort the berries as picked into two grades, and always use new, clean boxes or baskets made of the whitest wood possible.

TEACH YOUR PICKERS NOT TO REST CARRIERS ON PLANTS

Fill rounding full with fruit of uniform quality all the way through. After it is picked keep it away from the air as much as possible. Fruit, if dry cooled, will keep much longer and keep fresher if kept in tight crates. Ventilation in crates and baskets does more harm than good; to prove this, pick a basket of nice berries, put in a shady but airy place, and at the end of twenty-four hours the only bright and good berries will be in the bottom of the basket away from ventilation and light.''

Various strawberry hints : Strawberries should be picked or nipped off the vines with the stems on and not pulled off without the hulls. If picked with short stems a better appearance is given them and they stand shipment better. The patch should be picked regularly, clean, and often—once a day on picking days. See that the pickers do not crush or injure the vines, nor break off fruit clusters. Allow no green or overripe berries in the boxes. Permit no berries to rot on the vines ; a rotten berry on a cluster retards the development of other berries on that cluster. Pick one side of a row, and then the other side ; this does not apply to strawberries in hills or single hedge-rows, but for all wider rows this is the only method that insures clean picking. For near-by market only berries which are *fully colored* should be picked ; for distant market they may be picked a little sooner.

In picking strawberries, says a western grower, do not allow the pickers to touch the berries at all, but handle them by the stem and lay them in the boxes one by one as they are picked. Pick every ripe berry in the patch every day. Be honest. Do not allow pickers to put any trashy, rotten or green berries in the box. To avoid this it is absolutely necessary to have a superintendent in the patch and directly among the pickers.

A Wisconsin grower writes : Pickers paid by the day are most profitable ; they pick better, spoil less fruit and are more satisfactory.

A Massachusetts man says : We pay two cents a quart for careful picking.

Pickers should never be allowed to walk over the beds or handle berries except by the stem, which should be pinched off one-half to three-quarters of an inch from the berry and the berries carefully placed in the boxes. Good superintendence in the field is better than sorting and packing in the packing house. —H. E. M., Mississippi.

Be careful to have the berries clean and as uniform in size as possible. —A. W., Ohio.

In topping the basket the berries should be all

turned with the stem down and point up.   It makes
the fruit more attractive and commands better
prices.—E. W. R.

Build packing shed close to the patch.   Have an

overseer to every twenty to
forty pickers. Use carriers
containing six boxes. —
B. B., Illinois.

Have your baskets and
crates neat and clean; fill
baskets so they will go
on the market slightly
rounded.   A few fresh
leaves laid on the top of

"STEMS DOWN AND POINTS UP"   the boxes sometimes add
to their attractiveness.   Do not hide all the big
berries, but be sure they do not all come on top.—
E. W., New York.

The pickers sort the berries, putting the small,
soft or otherwise inferior fruit in one basket, while the
rest are put in the other baskets.   The pickers
arrange the berries neatly on the top of each basket,
thus presenting a neat appearance.   The culls, or
seconds, are sold to peddlers to do with as they
choose.—E. C. TICE.

Sandy, gritty berries find a poor market.   Keep
the fruit clean by mulching.—E. W. A.

We pick our berries every day in the berry
season ; there is no other way to do it.   You can not
pick a strawberry that is two days old and send it to
a distant market.   It must be picked when it is
exactly at the right stage for picking, and if you take

care to do that, you can ship them 1,000 miles if you want to.—PARKER EARLE, Illinois.

Strawberry packages : Different kinds of packages are used,—some localities prefer the sixteen-quart gift crate, some the twenty-four-quart gift crate, and some the thirty-two-quart gift crate. These crates are made of wood ends, and wood-veneer sides, tops and bottoms ; they are usually purchased in the flat and the buyer nails the different parts together during leisure hours. The boxes for the crates are generally made of thin wood-veneer; paper boxes are on the market, but have not as yet become very popular. Some of the boxes are square, and some are made like small, oblong baskets without handles. A crate, filled with boxes or baskets, costs but a trifling sum.

A GROUP OF STRAWBERRY PICKERS;
SHOWING ONE KIND OF CARRIER AND
SHIPPING CRATE

In a few localities the " return crate " is still in favor. These are more substantially made, and are supposed to be returned free (minus boxes) by the express company. This style of crate usually contains forty-eight or sixty quarts.

Consult your market man in regard to the best kind of package to use in your locality, and follow his advice.

RASPBERRIES AND BLACKBERRIES.—These are

not picked with the stems on ; nor need they as a rule be picked every day. Otherwise, many of the foregoing strawberry suggestions will apply equally well here.

Blackberries must not be allowed to get too ripe, or they will not "hold up" in shipping. And this fruit turns to a dingy, reddish color, after picking, if not kept out of the sun.

Instruct your pickers to handle raspberries and blackberries as if they were made of fragile glass that would be crushed by the slightest pressure.

Where black raspberries are grown in quantities for factories that evaporate or dry the product, picking is sometimes done by mechanical harvesters. Farmers' Bulletin No. 213 tells all about this method; write to the Secretary, U. S. Department of Agriculture, Washington, D. C., and ask for a copy. Black raspberries for table use are always hand picked ; red raspberries and blackberries, on account of their softness, are not adapted to the operation of mechanical harvesters.

Packages : Black raspberries and blackberries are usually packed in the same way as strawberries. Red raspberries, however, are almost always packed in *pint* boxes ; this delicate fruit is not suited to the larger quart boxes.

DEWBERRIES.—In growing dewberries on a large scale one of the serious problems is that of securing pickers, says O. B. Whipple, Colorado Agricultural Experiment Station. The average picker will pick from five to seven crates a day, and this means that it will take from eight to ten average pickers to pick an acre per day. The general practise is to pick every third day, and the large patch may be divided so as to furnish the pickers employment every day.

The pickers must at least wear a glove on the hand used to lift the vines ; and most of them wear a glove with the tips of the fingers removed on the picking hand. The pickers should be made to grade the fruit, and the best way is to have them put the culls in certain boxes and pay them for picking these the same as first-class fruit. This plan provides a place for fruit the picker gathers and hates to throw away because it fills up. Dewberries should be picked when a full glossy black. Berries which have gone beyond this stage and turned a dull or ashy color are too ripe to ship. The cull box is the place for overripe, dry, and poorly-colored berries. Ripe berries start to mold if packed for shipment.

Dewberries should not be picked when moist, as after a heavy dew or rain. They must be kept out of the sun after being picked. Pickers in Colorado are paid by the twenty-four-pint crate, thirty cents, if they pick part of the season, and thirty five cents if they finish the season. If the grower does not protect himself in this way, some of the pickers will leave him when picking gets poor.

Packing : Ever since dewberries were first grown in Colorado, says Prof. Whipple, several styles of packages have been used, but the crate now commonly used comes as near perfection as any. This crate holds twenty-four pint veneer boxes, twelve in each deck. The general practise is for the pickers to sort the berries and then all the packer has to do is to see that the boxes are full and not overfull, and possibly throw out a few defective berries overlooked by careless pickers. When packed and covered the crates should be ricked up end to end, preferably under an open shed, and allowed thoroughly to air out before shipping. If possible, it is a good plan to let them air over night and ship in the morning; unless well aired out the fruit molds in transit. Shippers should also insist on the car being well ventilated; icing only seems to aggravate molding.

In North Carolina, says F. C. Reimer, dewberries are packed and shipped in thirty-two-quart crates.

In many other localities they are packed in regulation strawberry packages, — sixteen or twenty-four-quart crates.

CURRANTS AND GOOSEBERRIES. —Currants should be picked when fully matured and colored, but not when overripe. Of course they are picked with stems on, in clusters, just as they grow.

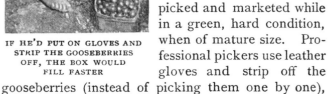

IF HE'D PUT ON GLOVES AND STRIP THE GOOSEBERRIES OFF, THE BOX WOULD FILL FASTER

Gooseberries are usually picked and marketed while in a green, hard condition, when of mature size. Professional pickers use leather gloves and strip off the gooseberries (instead of picking them one by one), and then run the berries through a fanning machine to blow out leaves, etc.

In some localities currants and gooseberrries are packed in quart boxes the same as strawberries ; consult your marketman.

GRAPES.—The picker should not be allowed to touch the bunches with the hand, but handle them by the stem. In packing, the clusters are lifted with thumb and finger, and with sharp-pointed grape scissors all green, imperfect or bruised berries are deftly removed. Thus the bloom on the grapes is preserved. None but perfectly sound baskets should be used ; green baskets cause mold. Where Concords have been picked in warm fall weather, all the cracked and bruised berries will show some mold, but as, in packing, all these are scrupulously removed, no harm is done. The Concord is never fit for shipping long distances, unless it has been first carefully picked, then wilted, and then packed. The packers soon learn to lay in the clusters so as to fill the baskets just level.

The grape improves little, if any, in flavor after picking, and ships best when fully ripe. The fruit should be allowed to hang on the vine as long as possible. The grape is best picked during the heat of the day as the stems are then least brittle and fewer berries are split or torn loose from the bunch. Picking after a rain and before the bunches have dried out is not advisable. The fruit should be handled in shallow boxes, or trays, and removed to the packing house for further wilting before packing. The common practise is to pick during the middle of the day, and pack this fruit out the following fore-noon. Whether or not this will be a long enough wilt will depend upon the ripeness of the fruit and its condition at packing time.

Concord grapes are preferably packed in eight-pound Climax baskets, the baskets to be well filled and faced up to hide all stems. Usually Delaware and Niagara grapes are packed in smaller-sized baskets. California grapes are shipped in square, open baskets packed in crates.

PLATE XV.

COLUMBUS

HOUGHTON

CHAUTAUQUA

PEARL

DOWNING

Chapter XIV

MARKETING

Establish a reputation for honest, uniform, high-grade products, and then live up to your brand.—Farmer Vincent.

There are various ways of selling berries after they are grown, and the best way for one grower may not be the best way for another. Generally speaking, it is a good idea to cater to the home market all you can, shipping the surplus, if any, to near-by towns. One can peddle, deliver to regular customers, or place the berries in the local groceries on commission. If the berries are nicely arranged in clean quart boxes, not putting the finest and largest ones on top and the small ones in the bottom, some grocer will be glad to get them, and his customers will willingly pay a few cents more for them than for inferior berries or those shipped from a distance.

If you live within reach of a cannery, it is sometimes advisable to sell your product for preserving purposes. Canneries are usually glad to make yearly contracts with reliable growers in the vicinity. Or it is possible to buy a home canning outfit and do your own preserving.

"With me advertising pays extra well in selling vegetables and small fruits," says a writer in American Cultivator. "I had a lot of large advertising cards, 12 x 15 inches, made at the printing office. When my first crop is ready for sale, which is rhubarb, I put a printed 'Rhubarb' card on a sign-board fastened to a post in the door-yard, where everybody can see it when passing. When the rhubarb season is over, I put up another card, 'Berries for Sale.' Next I put up 'Plums for Sale,' later 'Apples for Sale,' and so on. With this good way of advertising I sell large lots of fruit, vegetables, etc. It saves a large amount of labor in peddling the crop."

SHIPPING TO COMMISSION MEN.—"Get in touch with fruit dealers or commission men in good markets and get them to handle the fruit," says H. H. Hume. "Either ship to them on consignment or sell to them outright. Pick good, reliable men; send them good material, treat them squarely, and you will receive like treatment in return."

Yes, but everything depends upon picking "good, reliable men," for, as many fruit growers have found to their sorrow, all commission men are not reliable. Recently a little booklet has come to my desk, entitled, "Membership List, National League of Commission Merchants of the United States." The headquarters of this organization are at 202 Franklin Street, New York City, and the booklets are sent free to any one who asks for them. In the booklet I find this statement : " Our organization lays its foundation on the personal integrity and financial responsibility of its individual members "; then follows a list of members, and their addresses, in all the principal cities. Undoubtedly there are reliable commission firms that are not listed in the League, but unless I knew them I should prefer to take my chances on a listed firm.

Well-known shippers and commission men contribute the following hints, which are well worth remembering :

Don't ship to every strange house that solicits your consignment. Get a good solid house and stick to it.

Top-notch prices are governed very much by appearance. The eye of the buyer must be attracted.

TOP-NOTCH PRICES DEPEND LARGELY UPON APPEARANCE

In berry shipments, it is better policy to place the best berries in the bottom of the baskets than on top. The old trick of putting inferior berries below, and topping the baskets with choice fruit, has caused buyers to become suspicious.

The prices for shipments from a distance depend largely upon the manner of packing, as well as the style of package used.

If fruits are carefully assorted according to quality, size and appearance, the returns will more than offset the labor and time employed. This sorting, however, should never be done in the sun, but rather under a tree or shed, or in some cool, shady place.

It will pay the shipper of prime goods to label each package with his name and address, as well as the name of the farm. This not only creates a reputation, but greatly helps the commission man to make sales. There are two styles of labels used in marking fruit packages,—the ordinary stencil or stamp, and the one printed on paper to be pasted on the package. The former usually states the name of the variety, where and by whom grown. The latter, in addition to this, may be made up in colors and have a picture of the kind of produce for which it is to be used. Either style is good, but with conditions as they exist to-day, the neater and more attractive the label, the quicker it catches the eye of the public, and as a result the more ready the sale, particularly when the produce is in first-class condition. The one thing to be avoided in labeling any fruit or vegetable package is the placing of a label for first-class or fancy-grade produce on a package containing second-class or inferior grades. Practises of this kind will invariably result in a loss of both money and reputation.

It is best never to make a shipment (unless there is a previous understanding) without notifying the commission man. This gives him a good chance properly to dispose of the goods in quick time.

Co-operative Selling.—Now we come to what is, I think, one of the best of all marketing methods. Central packing houses, fruit-growers' exchanges, co-operative marketing associations, and similar organizations, are in successful operation in many parts of the country,—and the idea is spreading. These concerns are usually incorporated, and the surrounding fruit growers own stock and of course control the management.

A Maryland exchange : One of the best organized and managed of the co-operative selling associations that it has been my pleasure to look into, says Prof. W. N. Hutt, is the Peninsula Produce Exchange of the Eastern Shore of Maryland. It has twenty-five local shipping points, at each of which is an agent who inspects and brands the grade of produce, and reports to the head office at Olney the amounts and grades of fruit and truck received. The general manager in the head office is in touch by wire with prices in all the large markets, and as soon as the daily reports of receipts and grades are wired in from his local agents, he is in a position to make his sales and place his consignments where the demand is greatest. The exchange spends more than $10,000 annually in telegrams regarding crops, markets and prices. The capital stock of the exchange was reported in 1905 at $31,000. This was owned by the 2,500 farmers who sell through the exchange. In 1905 a dividend of seven per cent. was declared, and in 1906 a ten per cent. dividend. In addition to this a surplus was laid by for emergencies. The exchange forwards annually thousands of cars of both sweet and Irish potatoes in addition to other truck and fruit. It is reported as doing an annual business of about $2,000,000.

A co-operative selling association or exchange if properly organized and managed may be of immense value to the growers of fruit crops. It gives the small producer the privilege of shipping in car lots, which to-day is the only economic base for commercial fruit growing. It insures the advantages of a uniform grade of products. Buyers and large commission firms are willing to deal with a company, where they can not take the time and expense to hunt up the produce of the individual grower. Many associations save considerable money to their stock-holders by the purchase of baskets and fertilizers at wholesale rates.

Like any other stock company, the success of the enterprise depends largely on the loyalty of its stockholders and directors. A stockholder should not withhold his produce from the association when he can occasionally get a few cents better price elsewhere. In a season's business he would almost invariably do better to deal entirely through the organized channels of the company than to sell any of his stuff elsewhere. Local jealousies should be overcome and not allowed to impede business.

A good general manager in such a company is essential, and he should be paid what he is worth. A man who can manage a million-dollar business successfully, and make it pay a good dividend, can command the salary of a bank president and he should get it. Farmers' exchanges sometimes break up because farmer directors try to retain a $5,000 manager on a $1,000 salary.

Co-operative grape selling: During the past twelve years, writes S. S. Crissey, grape shipments from the Chautauqua-Erie district have ranged from

WHEN HAULING FRUIT, HAVE SPRINGS ON THE WAGON

a minimum in round numbers of 4,000 to a maximum of 8,000 cars. Every car of these grapes has been grown in nine towns bordering on Lake

Erie. Seven of these towns are in Chautauqua county, N. Y., and two are in Erie county, Pa. One of the most difficult problems in the history of the industry has been the safe, economical sale of the thousands of car-loads annually grown. For table use our grapes go in the eight-pound Climax basket. They go in car lots of 3,000 baskets each. The Grape Union does the selling for half a cent per basket. It has an inspector at the loading stations, and places its own salesmen in such leading markets as Chicago, Minneapolis, New York, etc. The Union advances five cents per basket and the balance at the close of the season. During the past five years there has been a rapid increase in the production of unfermented wines. The largest of these grape-juice establishments is at Westfield, N. Y.

How they co-operate in Colorado : One of the strong points in favor of the association idea, as worked out in Colorado, says W. Paddock, is the possibility of a fairly uniform pack. This results in better prices, since buyers have the assurance that all associations strive to make their goods as nearly uniform as possible. Then, contrary to the idea often advanced that poor fruit brings as great a price as good, the most rigid grading must be practised, and the intention is to place each fruit in its proper grade ; thus only the best grade sells for the highest price, and, indeed, the grower of inferior fruit is fortunate to dispose of his crop at all. There are two methods of packing and grading fruit. In one instance, the association does all the packing, the growers delivering the fruit to the packing house just as it is picked.

Here the packers, under the direction of a superintendent, sort the fruit into the various grades, and at the same time pack it. Should there be any culls, they are returned to the grower and are at his disposal.

Each grower is given a number, which is used to designate his fruit throughout the season. As each crate or box is packed, it is marked with his number and the grade. When the packages are loaded into the cars, the number of packages, the varieties and the various grades which belong to any grower, are kept account of and duly recorded. In this way the price for each package of fruit in any car is easily determined.

But where there is a very large amount of fruit to be handled it is impossible for the association to do the packing, consequently the growers assume this work. With this arrangement, the association employs an inspector, whose duty it is to inspect each load as it is delivered. This he does by opening some of the packages. If the pack is satisfactory, not more than two may be opened. If unsatisfactory, several may be examined, and if all run under the inspector's standard, the entire load must either be placed in a lower grade or else be repacked. It will be seen that a great deal depends on the inspector, and that it is a difficult position to fill. Upon him depends the reputation of the association, so he must be entirely free to do the work as he sees fit. Each man's fruit is kept track of by numbers, as in the former case.

Most of the Colorado associations have now adopted the latter system, although nearly all have tried the former. The ideal method is, no doubt, to have all packing done at a central building, but a limit to the amount of fruit which can be handled is soon reached. It is found difficult in practise to keep track of a large number of packers at a central point, and careless work is the result. But when each grower looks after his own packing, he has a wholesome respect for the decision of the inspector. It is very

THIS BERRY GROWER COMBINES BUSINESS AND PLEASURE

expensive to repack a lot of fruit, and if he is obliged to do this or else let it be sold as a lower grade, it usually results in greater pains being taken in the future. But with the best of systems, poorly packed fruit will occasionally find its way to market.

The association charges a commission on all sales, usually five per cent. to defray expenses. Then, in case the packing is done by the association, an additional charge is made to cover the cost of the package and packing. Any fruit grower may become a member of the association so long as there is stock for sale, and the owner of one share is entitled to all its privileges. The number of shares one individual may own is limited.

Here are the articles of incorporation and the by-laws of a Colorado fruit-growers' association, taken from Colorado Experiment Station Bulletin No. 122, and perhaps some of you can make these (with any necessary changes) the basis for a selling organization in your own locality :

ARTICLES OF INCORPORATION

I.

The name of the said Association shall be the Grand Junction Fruit Growers' Association.

II.

The objects for which the said Association is created are to buy and sell fruit, vegetables, hogs, meat stock and all the products of Mesa county, both fresh and manufactured; to erect, operate and maintain canning and packing factories and commission houses; to manufacture any and all products of Mesa county; to lease, mortgage and sell said business, and to borrow money for carrying on the same, and to pledge their property and franchise for such purpose. To acquire by purchase, or otherwise, and own real estate, buildings, machinery and all the necessary power and power plants for carrying on said premises, and to lease, mortgage and sell the same.

III.

The term of existence of said Association shall be twenty years.

IV.

The capital stock of the said Association shall be twenty-five thousand dollars ($25,000), divided into five thousand (5,000) shares of five dollars ($5) each.

V.

The number of Directors of said Association shall be seven, and the names of those who shall manage the affairs of the Association for the first year of its existence are ———

VI.

The principal office of said Association shall be kept at Grand Junction in the said county, and the principal business of said Association shall be carried on in said county of Mesa.

VII.

The stock of said Association shall be non-assessable.

VIII.

The Directors shall have power to make such prudential By-Laws as they may deem proper for the management of the affairs of the Association not inconsistent with the laws of this state, for the purpose of carrying on all kinds of business within the objects and purposes of the Association.

BY-LAWS.

ARTICLE I.

Section 1. The Board of Directors provided for in the articles of incorporation of this Association, shall be elected annually at the regular annual meeting of the stockholders, as hereinafter provided, and shall hold their offices until their successors are elected and qualified.

Section 2. Said Directors shall be stockholders in said Association and shall be fruit growers in Grand Valley and shall be residents of Mesa county, Colorado.

Section 3. Any vacancy occurring in the Board of Directors shall be filled by the remaining members of the Board.

ARTICLE II.

Section 1. The Board of Directors shall, as soon as may be, after their election, elect a President and Vice–President from among their number, who shall hold their offices for one year, and at said meeting the said Board shall appoint a Secretary, Treasurer and Manager, who shall be subject to removal at any time.

Section 2. The Secretary, Treasurer and Manager shall each when required by the Board, give bond in such sum and with such security as the Directors may require, conditioned on the faithful performance of their duties, and turn over to their successors in office all books, papers, vouchers. money, funds and property of whatsoever kind or nature belonging to the Association, upon the expiration of their respective terms of office, or upon their being removed therefrom, or with such other conditions as may be proper.

Section 3. The President shall preside at all meetings of

the Directors or Stockholders. He shall sign as President all certificates of stock, and all other contracts and other instruments in writing, which may have been ordered by the Board of Directors.

Section 4. The Vice-President shall, in the absence of or disability of the President, perform his duties.

Section 5. The Manager shall have full charge of the commercial and shipping department of the Association. He shall receive all money arising from the sale of fruit and other commodities handled by the Association, and pay the same to the parties entitled thereto, and render a true account thereof; and he shall also be the Treasurer of this Association and safely keep all money belonging to the Association, and disburse the same under the direction of the Board of Directors, except as hereinabove set forth.

Section 6. The Secretary shall keep a record of the proceedings of the Board of Directors and also of the meetings of the Stockholders. He shall also keep a book of blank certificates of stock, fill up and countersign all certificates issued, and make the corresponding entries upon the marginal stub of each certificate issued. He shall keep a stock ledger in due form, showing the number of shares issued to and transferred by any stockholder, and date of issuance and transfer. He shall have charge of the corporate seal, and affix the same to all instruments requiring a seal. He shall keep in the manner prescribed by the Board of Directors, all accounts of the Association with its stockholders, in books provided for such purpose. He shall discharge such other duties as pertain to his office, and as may be prescribed by the Board of Directors.

Section 7. These By-Laws may be amended by the Board of Directors at any special meeting thereof, called for that purpose, a notice of such proposed amendment being given in the call for such special meeting.

ARTICLE III.

Section 1. The regular meetings of the Board of Directors shall be held at the office of the company, on the first (1st) day of each month, except when the first day comes on Sunday or legal holiday, then on the following day.

Special meetings of the Board of Directors may be called by the President when he may deem it expedient or necessary, or by the Secretary, upon the request of any three members of said Board.

Section 2. A majority of the Board of Directors shall constitute a quorum for the transaction of business, but a less number may adjourn from day to day upon giving notice to absent members of the said Board, of such adjournment.

Section 3. The Board of Directors shall have power:

First—To call special meetings of the stockholders whenever they deem it necessary, by publishing a notice of such meeting once a week for two weeks next preceding such meeting in some newspaper published in Grand Junction, Colorado.

Second—To appoint and remove at pleasure all employees and agents of the Association, prescribe their duties, where the same have not been prescribed by the By-Laws of the Association, fix their compensation, and when they deem it necessary, require security for the faithful performance of their respective duties.

Third—To make such rules and regulations not inconsistent with the laws of the state of Colorado, and Articles of Incorporation, or the By-Laws of the Association, for the guidance of the officers and the management of the affairs of the Association.

Fourth—To incur such indebtedness as they may deem necessary for carrying out the objects and purposes of the Association, and to authorize the President and Secretary to make the note of the Association, with which to raise money to pay such indebtedness.

Section 4. It shall be the duty of the Board of Directors:

First—To be caused to be kept a complete record of all their meetings and acts, and also the proceedings of the stockholders, present full statements at the regular annual meetings of the stockholders, showing in detail the assets and liabilities of the Association, and the condition of its affairs in general.

Second—To supervise all acts of the officers and employees, and require the Secretary, Treasurer and Manager to keep full and accurate books of account of their respective business.

ARTICLE IV.

Section 1. At the regular meeting in the month of January of each year, the Directors shall declare such dividends upon the capital stock, to all the stockholders then appearing of record, as may be warranted by the net earnings of the Association for the preceding year.

ARTICLE V.

Section 1. The Board of Directors may, whenever they shall deem it necessary, place on sale so much of the capital stock of the Association as may be necessary to raise funds, for the purpose of carrying out the objects and purposes of the organization of the Association, such stock to be sold only upon the following conditions :

First—That not more than three hundred (300) shares thereof be sold to any one person, firm or association of persons.

Second—That such stock be sold only to fruit growers in Grand Valley.

Third—That such stock be sold at not less than par value of Five Dollars ($5) per share.

ARTICLE VI.

Section 1. The annual meeting of the stockholders for the election of Directors, shall be held on the third (3d) Saturday in January of each year, but if, for any reason, it should not be held on such day, it may then be held on any day subsequent thereto, as hereinafter provided.

Section 2. The Board of Directors shall be elected by the Stockholders at the regular annual meeting. Public notice of the time and place of holding such annual meeting and election, shall be published not less than ten (10) days previous thereto, in some newspaper of general circulation printed in Grand Junction, and the said election shall be made by such of the stockholders as shall attend for that purpose, either in person or by proxy, provided a majority of the outstanding stock is represented. If a majority of the outstanding stock shall not be represented, such meeting may be adjourned by the stockholders present for a period not exceeding sixty (60) days. All elections shall be by ballot, and each stockholder shall be entitled to as many votes as he or she owns shares of stock in said Association ; provided, however, that no person who is not himself a stockholder shall be allowed to represent by proxy any stockholder in the said Association.

The persons receiving the greatest number of votes shall be the Directors for the ensuing year, and until their successors are elected and qualified.

ARTICLE VII.

Section 1. Certificates of stock may be transferred at any time by the holders thereof, or by attorney in fact or legal representative. Such transfer shall be made by endorsement on the certificate of stock and surrender of the same ; provided, such transfer shall not be valid until the same shall have been noted in the proper form on the books of the Association.

The surrendered certificates shall be cancelled before a new certificate in lieu thereof shall be issued, and no transfer of any share of stock shall be valid or allowed upon the books of the Association upon which any deferred payments are due and unpaid, nor which has not been sold and transferred in accordance with the provisions of the By-Laws of the Association.

Section 2. Any stockholder desiring to dispose of his stock in said Association, shall deposit the same with the Secretary of the Association, and the same shall be sold by the said Secretary at not less than par for account of such stockholder, within sixty (60) days from date of such deposit, under the restriction of Section 1, Article 5, of these By-Laws; provided, that if the Secretary shall not have sold such stock at the expiration of sixty (60) days, then such stock may be returned to such stockholder, and be disposed of by him, without restriction or limitation by the Association.

ARTICLE VIII.

Section 1. All members of this Association are required to market all their fruit through the Association and bear their proportionate share of the expenses of handling the same.

Section 2. Any member may have the privilege of selling his own fruit at the orchard, but no sales of fruit shall be made to a dealer in fruit, or to any person who buys to ship outside the county. In case of the sale of the entire crop of any particular fruit or fruits, by reporting the same to the Association, one-half (½) only of the regular commission will be charged.

Section 3. Any member having any grievance or cause of complaint as to treatment of his fruit by the Association, can appeal to the Board of Directors, whose decision shall be final.

Section 4. All members must pack their fruit for shipping in a neat and workman-like manner, and pack the same in standard-size packages, as adopted and in general use by the Association, having placed thereon their name or number.

ARTICLE IX.

Section 1. A purchaser of stock in this, the Grand Junction Fruit Growers' Association, shall hereafter receive of the profits of the Association, in proportion to the money he has invested.

INDEX